Image Acquisition and Processing with LabVIEW™

IMAGE PROCESSING SERIES

Series Editor: Phillip A. Laplante, Pennsylvania Institute of Technology

Published Titles

Adaptive Image Processing: A Computational Intelligence Perspective
Stuart William Perry, Hau-San Wong, and Ling Guan

Image Acquisition and Processing with LabVIEW™
Christopher G. Relf

Image and Video Compression for Multimedia Engineering
Yun Q. Shi and Huiyang Sun

Multimedia Image and Video Processing
Ling Guan, S.Y. Kung, and Jan Larsen

Shape Analysis and Classification: Theory and Practice
Luciano da Fontoura Costa and Roberto Marcondes Cesar Jr.

Software Engineering for Image Processing Systems
Phillip A. Laplante

Image Acquisition and Processing with LabVIEW™

Christopher G. Relf

CRC PRESS

Boca Raton   London   New York   Washington, D.C.

Library of Congress Cataloging-in-Publication Data

Relf, Christopher G.
    Image acquisition and processing with LabVIEW / Christopher G. Relf
        p.   cm. (Image processing series)
    Includes bibliographical references and index.
    ISBN 0-8493-1480-1
    1. Image processing--Digital techniques. 2. Engineering instruments--Data processing. 3.
    LabVIEW. I. Title. II. Series.
TA1632.R44 2003
621.36'7—dc21   2003046135
                                                                                    CIP

Visit the CRC Press Web site at www.crcpress.com

© 2004 by CRC Press LLC

No claim to original U.S. Government works
International Standard Book Number 0-8493-1480-1
Library of Congress Card Number 2003046135

Foreword

The introduction of LabVIEW over 16 years ago triggered the Virtual Instrumentation revolution that is still growing rapidly today. The tremendous advances in personal computers and consumer electronics continue to fuel this growth. For the same cost, today's computers are about 100 times better than the machines of the LabVIEW 1 days, in CPU clock rate, RAM size, bus speed and disk size. This trend will likely continue for another 5 to 10 years.

Virtual Instrumentation first brought the connection of electronic instruments to computers, then later added the ability to plug measurement devices directly into the computer. Then, almost 7 years ago, National Instruments expanded the vision of virtual instrumentation when it introduced its first image acquisition hardware along with the LabVIEW Image Analysis library. At the time, image processing on a personal computer was still a novelty requiring the most powerful machines and a lot of specialized knowledge on the part of the system developer. Since then, computer performance and memory size have continued to increase to the point where image processing is now practical on most modern PCs. In addition, the range of product offerings has expanded and higher-level software, such as Vision Builder, has become available to make development of image processing applications much easier.

Today, image processing is fast becoming a mainstream component of Virtual Instrumentation. Very few engineers, however, have had experience with image processing or the lighting techniques required to capture images that can be processed quickly and accurately. Hence the need for a book like this one. Christopher Relf has written a very readable and enjoyable introduction to image processing, with clear and straightforward examples to illustrate the concepts, good references to more detailed information, and many real-world solutions to show the breadth of vision applications that are possible. The lucid (pun intended) description of the role of, and options for, lighting is itself worth the price of the book.

Jeff Kodosky
National Instruments Fellow
Co-inventor of LabView

Preface

Image Acquisition and Processing with LabVIEW fills a hole in the LabVIEW technical publication range. It is intended for competent LabVIEW programmers, as a general training manual for those new to National Instruments' (NI) Vision application development and a reference for more-experienced vision programmers. It is assumed that readers have attained programming knowledge comparable to that taught in the NI LabVIEW Basics II course (see http://www.ni.com/training for a detailed course outline). The book covers introductions and theory of general image acquisition and processing topics, while providing more in-depth discussions and examples of specific NI Vision tools.

This book is a comprehensive IMAQ and Vision resource combining reference material, theory on image processing techniques, information on how LabVIEW and the NI Vision toolkit handle each technique, examples of each of their uses and real-world case studies, all in one book.

This is not a "laboratory-style" book, and hence does not contain exercises for the reader to complete. Instead, the several coding examples, as referenced in the text, are included on an accompanying CD-ROM in the back of the book.

The information contained in this book refers generally to the National Instruments Vision Toolkit version 6.1 (Figure 1). Several of the techniques explained herein may be perfectly functional using previous or future versions of the Vision Toolkit

A glossary has also been compiled to define subject-related words and acronyms.

THE COMPANION CD-ROM

Like most modern computer-related technical books, this one is accompanied by a companion CD-ROM that contains libraries of example images and code, as referenced in the text. Every wiring diagram shown in this book has corresponding source code on the CD-ROM for your convenience — just look in the respective chapter's folder for a file with the same name as the image's caption (Figure 2).

To use the code, you will need:

- LabVIEW 6.1 (or higher)
- LabVIEW Vision Toolkit 6.1 (or higher)
- An operating system that supports both of these components

Some of the examples may also require the following components to be installed:

- IMAQ
- Vision Builder 6.1 (or higher)
- IMAQ OCR Toolkit
- IMAQ 1394 Drivers

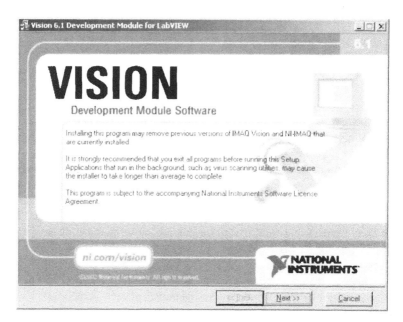

FIGURE 1 Vision 6.1

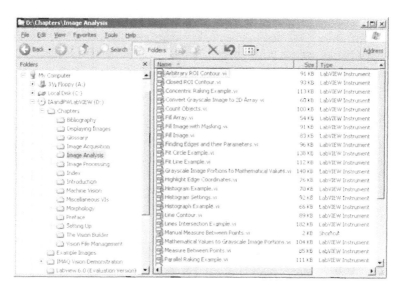

FIGURE 2 Companion CD-ROM Contents

The CD-ROM also contains all of the example images used to test the code featured in the book, a demonstration version of National Instruments LabVIEW 6.0, and a National Instruments IMAQ Demonstration that guides you through some of the features of NI-IMAQ.

Author

Christopher G. Relf is an Industrial Automation Software Engineering consultant and LabVIEW specialist. Previously, his work at JDS Uniphase Pty Ltd (www.jdsu.com) included the automation of several complex processes. including the laser writing of optical FBGs (Fiber Bragg Gratings) and their measurement. Mr. Relf was part of a strong LabVIEW team that designed and implemented a plug-in-style software suite that introduced an "any product, any process, anywhere" paradigm to both the production and R&D phases of product creation. As a Computational Automation Scientist with the Division of Telecommunications and Industrial Physics, CSIRO, (www.tip.csiro.au), Australia's premier scientific and industrial research organization, he was the principal software engineer of several projects including the automation of thin film filtered arc deposition and metal shell eccentricity systems. He has consulted to the New South Wales Institute of Sport (www.nswis.com.au) and provided advice on the development of automated sport science data acquisition systems, used to test and train some of Australia's premier athletes.

Mr. Relf completed his undergraduate science degree in applied physics at the University of Technology, Sydney, where he first learned the virtues of LabVIEW version 3, and has been a strong G programming advocate ever since. He gained his Certified LabVIEW Developer qualification from National Instruments in 2002.

Mr. Relf can be contacted via e-mail at Christopher.Relf@mBox.com.au

Acknowledgments

My gratitude goes out to all of the people at CRC Press LLC who have held my hand through the production of this book. Specifically, thanks to Nora Konopka, my Acquisitions Editor, who showed immense faith in taking on and internally promoting this project; Helena Redshaw, my Editorial Project Development Supervisor, who kept me on the right track during the book's development; and Sylvia Wood, my Project Editor, who worked closely with me to convert my ramblings into comprehensible format.

The user solutions featured were written by some very talented LabVIEW and Vision specialists from around the world, most of whom I found lurking either on the Info-LabVIEW Mailing List (http://www.info-labview.org), or the NI-Zone Discussion Forum (http://www.zone.ni.com). Theoretical descriptions of the Vision capabilities of LabVIEW are all very well, but it is only when you build a program that ties the individual functionalities together to form a useful application that you realize the true value of their use. I hope their stories help you to see the big picture (pun intended), and to realize that complete solutions can be created around core Vision components. Thank you for the stories and examples showing some of the real-world applications achievable using the techniques covered in the book.

A big thank you goes out to my two very talented reviewers, Edward Lipnicki and Peter Badcock, whose knowledge of theory often overshadowed my scientific assumptions — this book would not be anywhere near as comprehensive without your help. Thanks also to my proofreaders, Adam Batten, Paul Conroy, Archie Garcia, Walter Kalceff, James McDonald, Andrew Parkinson, Paul Relf, Anthony Rochford, Nestor Sanchez, Gil Smith, Glen Trudgett and Michael Wallace, who helped me correct most (hopefully) of the mistakes I made in initial manuscripts.

The section on application lighting is based primarily on information and images I gathered from NER (a Robotic Vision Systems Incorporated company) — thank you to Greg Dwyer from the Marketing Communications group for assisting me in putting it all together. Further information regarding NER's extensive range of industrial lighting products can be found at their Web site (http://www.nerlite.com).

I could not have dreamed of writing this book without the assistance of my local National Instruments staff, Jeremy Carter (Australia and New Zealand Branch Manager) and Alex Gouliaev (Field Sales Engineer). Your product-related assistance and depth of knowledge of NI products and related technologies were invaluable and nowhere short of astounding.

My apologies to my family, my friends and all of my English teachers. This book is pitched primarily to readers in the United States, so I have had to abandon many of the strict rules of English spelling and grammar that you tried so hard to instill in me. I hope that you can forgive me.

Thanks must also go to Ed, Gleno, Jase, Rocko and Adrian, who were always more than willing to help me relax with a few schooners between (and occasionally during) chapters.

Those who have written a similar-level technical book understand the draining exercise that it is, and my love and humble appreciation go to my family and friends, who supported me in many ways through this project. Ma and Paul, who planted the "you-can-do-anything" seed, and Ruth, who tends the resulting tree every day: I couldn't have done it without you all.

Christopher G. Relf

Dedication

To Don "old fella" Price.
I appreciate your inspiration, support, friendship
and, above all else, your good humor.

Introduction

Of the five senses we use daily, we rely most on our sense of sight. Our eyesight provides 80% of the information we absorb during daylight hours, and it is no wonder, as nearly three quarters of the sensory receptor cells in our bodies are located in the back of our eyes, at the retinas. Many of the decisions we make daily are based on what we can see, and how we in turn interpret that data — an action as common as driving a car is only possible for those who have eyesight.

As computer-controlled robotics play an ever-increasing role in factory production, scientific, medical and safety fields, surely one of the most important and intuitive advancements we can exploit is the ability to acquire, process and make decisions based on image data.

LabVIEW is a popular measurement and automation programming language, developed by National Instruments (NI). Initially aimed squarely at scientists and technicians to assist in simple laboratory automation and data acquisition, LabVIEW has grown steadily to become a complete programming language in its own right, with add-on toolkits that cover anything from Internet connectivity and database access to fuzzy logic and image processing. The image-based toolkit (called *Vision*) is particularly popular, simplifying the often-complex task of not only downloading appropriate quality images into a computer, but also making multifarious processing tasks much easier than many other packages and languages.

This book is intended for competent LabVIEW programmers, as a general training manual for those new to NI Vision application development and a reference for more-experienced Vision programmers. It is assumed that you have attained programming knowledge comparable to that taught in the NI LabVIEW Basics II course.

I sincerely hope that you will learn much from this book, that it will aid you in developing and including Vision components confidently in your applications, thus harnessing one of the most complex, yet common data sources — the photon.

Contents

1 Image Types and File Management

Files provide an opportunity to store and retrieve image information using media such as hard drives, CD-ROMs and floppy disks. Storing files also allows us a method of transferring image data from one computer to another, whether by first saving it to storage media and physically transferring the file to another PC, or by e-mailing the files to a remote location.

There are many different image file types, all with advantages and disadvantages, including compression techniques that make traditionally large image file footprints smaller, to the number of colors that the file type will permit. Before we learn about the individual image file types that the Vision Toolkit can work with, let us consider the image types themselves.

1.1 TYPES OF IMAGES

The Vision Toolkit can read and manipulate *raster* images. A raster image is broken into cells called *pixels*, where each pixel contains the color or grayscale intensity information for its respective spatial position in the image (Figure 1.1).

Machine vision cameras acquire images in raster format as a representation of the light falling on to a charged-coupled device (CCD; see Chapter 2 for information regarding image acquisition hardware). This information is then transmitted to the computer through a standard data bus or frame grabber.

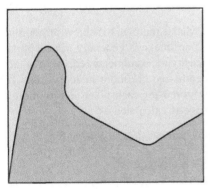
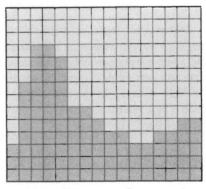

Image   Pixelated Raster Image Representation

FIGURE 1.1 Raster image.

An image's type does not necessarily define its image file type, or vice versa. Some image types work well with certain image file types, and so it is important to understand the data format of an image before selecting the file type to be used for storage. The Vision Toolkit is designed to process three image types:

1. Grayscale
2. Color
3. Complex

1.1.1 GRAYSCALE

Grayscale images are the simplest to consider, and are the type most frequently demonstrated in this book. Grayscale images consist of x and y spatial coordinates and their respective intensity values. Grayscale images can be thought of as surface graphs, with the z axis representing the intensity of light. As you can see from Figure 1.2, the brighter areas in the image represent higher z-axis values. The surface plot has also been artificially shaded to further represent the intensity data.

The image's intensity data is represented by its *depth*, which is the range of intensities that can be represented per pixel. For a bit depth of x, the image is said to have a depth of 2^x, meaning that each pixel can have an intensity value of 2^x levels. The Vision Toolkit can manipulate grayscale images with the following depths:

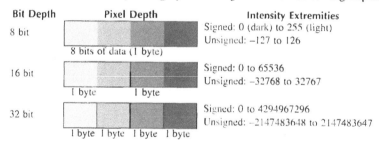

Bit Depth	Pixel Depth	Intensity Extremities
8 bit	8 bits of data (1 byte)	Signed: 0 (dark) to 255 (light) Unsigned: −127 to 126
16 bit	1 byte 1 byte	Signed: 0 to 65536 Unsigned: −32768 to 32767
32 bit	1 byte 1 byte 1 byte 1 byte	Signed: 0 to 4294967296 Unsigned: −2147483648 to 2147483647

Differing bit depths exist as a course of what is required to achieve an appropriate imaging solution. Searching for features in an image is generally achievable using 8-bit images, whereas making accurate intensity measurements requires a higher bit depth. As you might expect, higher bit depths require more memory (both RAM and fixed storage), as the intensity values stored for each pixel require more data space. The memory required for a raw image is calculated as:

$$MemoryRequired = Resolution_x \times Resolution_y \times BitDepth$$

For example, a 1024 × 768 8-bit grayscale would require:

$$MemoryRequired = 1024 \times 768 \times 8$$
$$= 6291456 Bits$$
$$= 786432 Bytes$$
$$= 768 kBytes$$

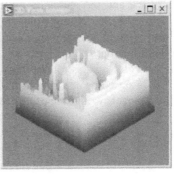

FIGURE 1.2 Image data represented as a surface plot.

Increasing the bit depth of the image to 16 bits also increases the amount of memory required to store the image in a raw format:

$$MemoryRequired = 1024 \times 768 \times 16$$
$$= 12582912 Bits$$
$$= 1572864 Bytes$$
$$= 1.536 MBytes$$

You should therefore select an image depth that corresponds to the next-highest level above the level required, e.g., if you need to use a 7-bit image, select 8 bits, but not 16 bits.

Images with bit depths that do not correspond to those listed in the previous table (e.g., 1, 2 and 4-bit images) are converted to the next-highest acceptable bit depths when they are initially loaded.

1.1.2 COLOR

Color images are represented using either the Red-Green-Blue (RGB) or Hue-Saturation-Luminance (HSL) models. The Vision Toolkit accepts 32-bit color images that use either of these models, as four 8-bit channels:

Color Model	Pixel Depth				Channel Intensity Extremities
RGB	α	Red	Green	Blue	0 to 255
HSL	α	Hue	Saturation	Luminance	0 to 255

The alpha (α) component describes the opacity of an image, with zero representing a clear pixel and 255 representing a fully opaque pixel. This enables an image to be rendered over another image, with some of the underlying image showing through. When combining images, alpha-based pixel formats have several advantages over color-keyed formats, including support for shapes with soft or antialiased edges, and the ability to paste the image over a background with the foreground information seemingly blending in. Generally, this form of transparency is of little use to the industrial vision system user, so the Vision Toolkit ignores all forms of α information.

The size of color images follows the same relationship as grayscale images, so a 1024 × 768 24-bit color image (which equates to a 32-bit image including 8 bits for the alpha channel) would require the following amount of memory:

$$
\begin{aligned}
MemoryRequired &= 1024 \times 768 \times 32 \\
&= 25165824 \text{ Bits} \\
&= 3145728 \text{ Bytes} \\
&= 3.072 \text{ MBytes}
\end{aligned}
$$

1.1.3 COMPLEX

Complex images derive their name from the fact that their representation includes real and complex components. Complex image pixels are stored as 64-bit floating-point numbers, which are constructed with 32-bit real and 32-bit imaginary parts.

Pixel Depth		Channel Intensity Extremities
Real (4 bytes)	Imaginary (4 bytes)	−2147483648 to 2147483647

A complex image contains frequency information representing a grayscale image, and therefore can be useful when you need to apply frequency domain processes to the image data. Complex images are created by performing a fast Fourier transform (FFT) on a grayscale image, and can be converted back to their original state by applying an inverse FFT. Magnitude and phase relationships can be easily extracted from complex images.

1.2 FILE TYPES

Some of the earliest image file types consisted of ASCII text-delimited strings, with each delimiter separating relative pixel intensities. Consider the following example:

```
0    0    0    0    0    0    0    0    0    0    0
0    0    0    125  125  125  125  125  0    0    0
0    0    125  25   25   25   25   25   125  0    0
0    125  25   25   255  25   255  25   25   125  0
0    125  25   25   25   25   25   25   25   125  0
0    125  25   25   25   255  25   25   25   125  0
0    125  25   255  25   25   25   255  25   125  0
0    125  5    25   255  255  255  25   25   125  0
0    0    125  25   25   25   25   25   125  0    0
0    0    0    125  125  125  125  125  0    0    0
0    0    0    0    0    0    0    0    0    0    0
```

At first look, these numbers may seem random, but when their respective pixel intensities are superimposed under them, an image forms:

0	0	0	0	0	0	0	0	0	0	0
0	0	0	125	125	125	125	125	0	0	0
0	0	125	25	25	25	25	25	125	0	0
0	125	25	25	255	25	255	25	25	125	0
0	125	25	25	25	25	25	25	125	0	
0	125	25	25	25	255	25	25	25	125	0
0	125	25	255	25	25	25	255	25	125	0
0	125	25	25	255	255	255	25	25	125	0
0	0	125	25	25	25	25	25	125	0	0
0	0	0	125	125	125	125	125	0	0	0
0	0	0	0	0	0	0	0	0	0	0

Synonymous with the tab-delimited spreadsheet file, this type of image file is easy to understand, and hence simple to manipulate. Unfortunately, files that use this structure tend to be large and, consequently, mathematical manipulations based on them are slow. The size of this example (11 × 11 pixels with 2-bit color) would yield a dataset size of:

$$MemoryRequired = 11 \times 11 \times 2$$

$$= 242 Bits$$

$$\approx 30 Bytes + delimter\ character\ bytes$$

In the current era of cheap RAM and when hard drives are commonly available in the order of hundreds of gigabytes, 30 bytes may not seem excessive, but storing the intensity information for each pixel individually is quite inefficient.

1.2.1 MODERN FILE FORMATS

Image files can be very large, and they subsequently require excessive amounts of hard drive and RAM space, more disk access, long transfer times, and slow image manipulation. Compression is the process by which data is reduced to a form that minimizes the space required for storage and the bandwidth required for transmittal, and can be either *lossy* or *lossless*. As its name suggests, lossless compression occurs

FIGURE 1.3 Lossy spatial compression example.

when the data file size is decreased, without the loss of information. Lossless compression routines scan the input image and calculate a more-efficient method of storing the data, without changing the data's accuracy. For example, a lossless compression routine may convert recurring patterns into short abbreviations, or represent the image based on the change of pixel intensities, rather than each intensity itself.

Lossy compression discards information based on a set of rules, ultimately degrading the data. For example, an image that is very large can often be resized to become a little smaller, effectively decreasing the amount of information, yet retaining the large features of the original image. Consider the example in Figure 1.3.

The smaller barcode image has been resized to 75% of the original image, but the resolution is still high enough to read the features required. Some lossy compression routines store color information at a lower resolution than the source image. For example, a photo-realistic image contains millions of colors, and a machine vision system may only need to find edges within the image. Often the number of colors used to represent the image can be dramatically decreased (some vision routines require images to be resampled to binary levels — black and white).

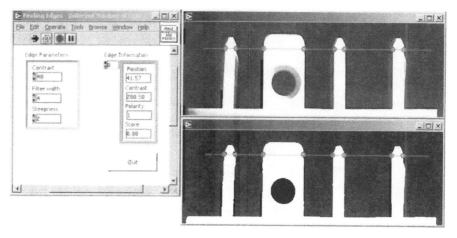

FIGURE 1.4 Finding edges: Different number of colors.

Consider the example in Figure 1.4. An edge detection routine has been executed on two similar images. The first is an 8-bit grayscale image (which contains 256 colors), and the second is a binary (2 colors) image. Although the amount of color data has been significantly reduced in the second image, the edge detection routine has completed successfully. In this example, decreasing the image's color depth to a binary level may actually improve the detection of edges; the change between background and object is from one extreme to the other, whereas the steps in the 256-color image are more gradual, making the detection of edges more difficult.

1.2.1.1 JPEG

Of the five image file formats that the Vision Toolkit supports by default, JPEG (Joint Photographic Experts Group) files are probably the most common. The JPEG format is optimized for photographs and similar continuous tone images that contain a large number of colors, and can achieve astonishing compression ratios, even while maintaining a high image quality. The JPEG compression technique analyzes images, removes data that is difficult for the human eye to distinguish, and stores the resulting data as a 24-bit color image. The level of compression used in the JPEG conversion is definable, and photographs that have been saved with a JPEG compression level of up to 15 are often difficult to distinguish from their source images, even at high magnification.

1.2.1.2 TIFF

Tagged Image File Format (TIFF) files are very flexible, as the routine used to compress the image file is stored within the file itself. Although this suggests that TIFF files can undergo lossy compression, most applications that use the format use only lossless algorithms. Compressing TIFF images with the Lempel-Zev-Welch (LZW) algorithm (created by Unisys), which requires special licensing, has recently become popular.

1.2.1.3 GIF

GIF (CompuServe Graphics Interchange File) images use a similar LZW compression algorithm to that used within TIFF images, except the bytes are reversed and the string table is upside-down. All GIF files have a color palette, and some can be interlaced so that raster lines can appear as every four lines, then every eight lines, then every other line. This makes the use of GIFs in a Web page very attractive when slow download speeds may impede the viewing of an image. Compression using the GIF format creates a color table of 256 colors; therefore, if the image has fewer than 256 colors, GIF can render the image exactly. If the image contains more than 256 colors, the GIF algorithm approximates the colors in the image with the limited palette of 256 colors available. Conversely, if a source image contains less than 256 colors, its color table is expanded to cover the full 256 graduations, possibly resulting in a larger file size. The GIF compression process also uses repeating pixel colors to further compress the image. Consider the following row of a raster image:

125	125	125	125	125	125	126	126	126	126	$\Rightarrow$	6(125)4(126)

Instead of represented each pixel's intensity discreetly, a formula is generated that takes up much less space.

1.2.1.4 PNG

The Portable Network Graphics file format is also a lossless storage format that analyzes patterns within the image to compress the file. PNG is an excellent replacement for GIF images, and unlike GIF, is patent-free. PNG images can be indexed color, true color or grayscale, with color depths from 1 to 16 bits, and can support progressive display, so they are particularly suited to Web pages. Most image formats contain a header, a small section at the start of the file where information regarding the image, the application that created it, and other nonimage data is stored. The header that exists in a PNG file is programmatically editable using code developed by a National Instruments' Applications Engineer. You are able to store multiple strings (with a maximum length of 64 characters) at an unlimited number of indices. Using this technique, it is possible to "hide" any type of information in a PNG image, including text, movies and other images (this is the method used to store region of interest (ROI) information in a PNG file). The source code used to realize this technique is available at the National Instruments Developer Zone at http://www.zone.ni.com (enter "Read and Write Custom Strings in a PNG" in the Search box).

1.2.1.5 BMP

Bitmaps come in two varieties, OS/2 and Windows, although the latter is by far the most popular. BMP files are uncompressed, support both 8-bit grayscale and color, and can store calibration information about the physical size of the image along with the intensity data.

1.2.1.6 AIPD

AIPD is a National Instruments uncompressed format used internally by LabVIEW to store images of Floating Point, Complex and HSL, along with calibration information and other specifics.

1.2.1.7 Other Types

Although the formats listed in this chapter are the default types that the Vision Toolkit can read, several third-party libraries exist that extend the range of file types supported. One such library, called "The Image Toolbox," developed by George Zou, allows your vision code to open icon (.ico), Windows Metafile (.wmf), ZSoft Corporation Picture (.pcx) files and more. The Image Toolbox also contains other functions such as high-quality image resizing, color-grayscale and RGB-HSL conversion utilities. More information (including shareware downloads) about George Zou's work can be found at http://gtoolbox.yeah.net.

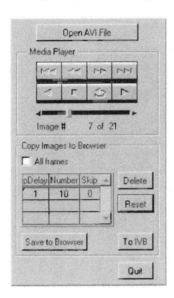

FIGURE 1.5 AVI external module for IMAQ vision builder.

Alliance Vision has released a LabVIEW utility that allows the extraction of images from an AVI movie file. The utility can be used as a stand-alone application, or can be tightly integrated into the National Instruments Vision Builder (Figure 1.5).

The utility also allows the creation of AVIs from a series of images, including movie compression codecs. More information about the Alliance Vision *AVI External Module for IMAQ Vision Builder* can be found at http://www.alliancevision.com.

1.3 WORKING WITH IMAGE FILES

Almost all machine vision applications work with image files. Whether the images are acquired, saved and postprocessed, or if images are acquired, processed and saved for quality assurance purposes, most systems require the saving and loading of images to and from a fixed storage device.

1.3.1 STANDARD IMAGE FILES

The simplest method of reading an image file from disk is to use IMAQ ReadFile. This Vision Toolkit VI can open and read standard image file types (BMP, TIFF, JPEG, PNG and AIPD), as well as more-obscure formats (Figure 1.6). Using IMAQ ReadFile with the standard image types is straightforward, as shown in Figure 1.7. Once you have created the image data space using IMAQ Create, the image file is opened and

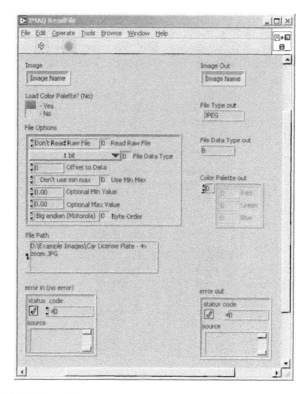

FIGURE 1.6 IMAQ ReadFile

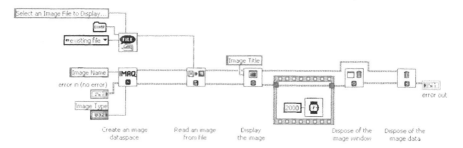

FIGURE 1.7 Loading a Standard Image File

read using IMAQ ReadFile. The image data is parsed and, if required, converted into the type specified in IMAQ Create. This image is then loaded into memory, and referenced by its *Image Out* output. Both *Image In* and *Image Out* clusters used throughout the Vision Toolkit are clusters of a string and numeric; they do not contain the actual image data, but are references (*or pointers*) to the data in memory.

LabVIEW often creates more than one copy of data that is passed into sub-VIs, sequences, loops and the like, so the technique of passing a pointer instead of the complete image dataset dramatically decreases the processing and memory required when completing even the simplest of Vision tasks.

Saving standard image files is also straightforward using IMAQ WriteFile, as shown in Figure 1.8. IMAQ WriteFile has the following inputs:

Input	Description
File type	Specify the standard format to use when writing the image to file. The supported values are: AIPD, BMP, JPEG, PNG, TIFF[a]
Color palette	Applies colors to a grayscale source image. The color palette is an array of clusters that can be constructed either programmatically, or determined using IMAQ GetPalette, and is composed of three-color planes (red, green and blue), each consisting of 256 elements. A specific color is achieved by applying a value between 0 and 255 for each of the planes (for example, light gray has high and identical values in each of the planes). If the selected image type requires a color palette and if one has not been supplied, a grayscale color palette is generated and written to the image file.

[a] IMAQ WriteFile has a cluster on its front panel where TIFF file-saving parameters can be set (including photometric and byte order options), although this input is not wired to the connector pane. If you use the TIFF format, you should wire the cluster to the connector pane, allowing programmatic access to these options.

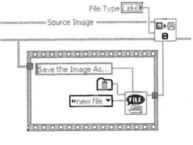

FIGURE 1.8 Saving a Standard Image File

1.3.2 Custom and Other Image Formats

Working with images files that are not included in the five standard types supported by IMAQ ReadFile can be achieved either by using a third-party toolkit, or decoding the file using IMAQ ReadFile's advanced features. Consider the example in Figure 1.9, which shows an image being saved in a custom file format. The size of the image (x,y) is first determined and saved as a file header, and then the image data is converted to an unsigned 8-bit array and saved as well. Opening the resulting file in an ASCII text editor shows the nature of the data (Figure 1.10)

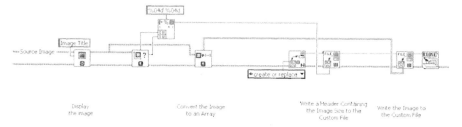

Display   Convert the Image   Write a Header Containing   Write the Image to
the image   to an Array   the Image Size to the   the Custom File
                                                    Custom File

FIGURE 1.9 Save an image in a custom format — wiring diagram.

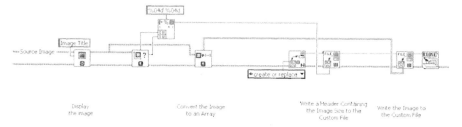

FIGURE 1.10 A custom file format image.

The first four characters in the file specify the x size of the image, followed by a delimiting space (this is not required, but has been included to demonstrate the technique), then the four characters that represent the y size of the image. When reading this custom file format, the header is first read and decoded, and then the remaining file data is read using IMAQ ReadFile (Figure 1.11).

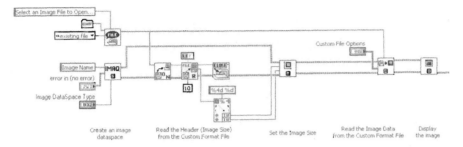

FIGURE 1.11 Load image from a custom format — wiring diagram.

When setting the file options cluster and reading custom files, it is vitally important to know the raw file's structure. In the previous example, the values in Figure 1.12 were used.

Input	Description
Read raw file	If set as *zero*, the remaining cluster options are ignored, and the routine attempts to automatically determine the type of standard image file (AIPD, BMP, JPEG, PNG and TIFF); otherwise the file is considered a custom format
File data type	Indicates how each pixel of the image is encoded; the following values are permitted: • 1 bit • 2 bits • 4 bits • 8 bits • 16 bits (unsigned) • 16 bits (signed) • 16 bits (RGB) • 24 bits (RGB) • 32 bits (unsigned) • 32 bits (signed) • 32 bits (RGB) • 32 bits (HSL) • 32 bits (float) • 48 bits (complex 2×24-bit integers)
Offset to data	The number of bytes to ignore at the start of the file before the image data begins; use this input when a header exists in the custom format file
Use min max	This input allows the user to set a minimum and maximum pixel intensity range when reading the image; the following settings are permitted: • Do not use min max (min and max are set automatically based on the extremes permitted for the *file data type*) • Use file values (minimum and maximum values are determined by first scanning the file, and using the min and max values found; the range between the min and max values are then linearly interpolated to create a color table) • Use optional values (use the values set in the *optional min value* and *optional max value* controls)
Byte order	Sets whether the byte weight is to be swapped (big or little endian; this setting is valid only for images with a file data type of 8 or more bits)

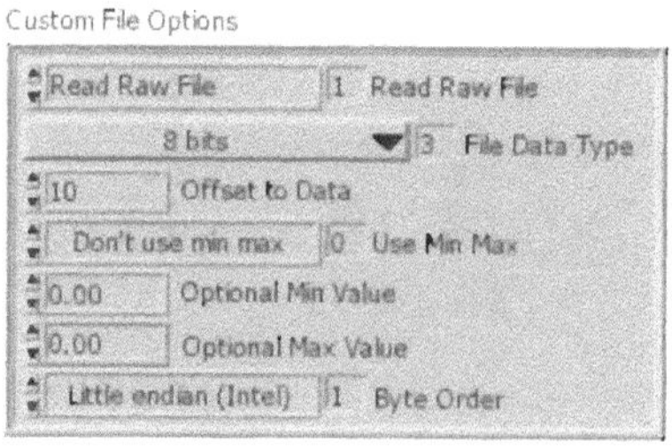

Custom File Options

Read Raw File   1   Read Raw File

          8 bits   ▼ 3   File Data Type

10   Offset to Data

Don't use min max   0   Use Min Max

0.00   Optional Min Value

0.00   Optional Max Value

Little endian (Intel) 1   Byte Order

FIGURE 1.12   IMAQ ReadFile - File Options

2 Setting Up

You should never underestimate the importance of your image acquisition hardware. Choosing the right combination can save hours of development time and strongly influence the performance of your built application. The number of cameras, acquisition cards and drivers is truly mind-boggling, and can be daunting for the first-time Vision developer. This book does not try to cover the full range of hardware available, but instead describes common product families most appropriate to the LabVIEW Vision developer.

2.1 CAMERAS

Your selection of camera is heavily dependent on your application. If you select an appropriate camera, lens and lighting setup, your efforts can then be focused (pun intended!) on developing your solution, rather than wrestling with poor image data; also, appropriate hardware selection can radically change the image processing that you may require, often saving processing time at execution.

An electronic camera contains a sensor that maps an array of incident photons (an optical image) into an electronic signal. Television broadcast cameras were originally based on expensive and often bulky Vidicon image tubes, but in 1970 Boyle invented the solid state charged-coupled device (CCD). A CCD is a light-sensitive, integrated circuit that stores irradiated image data in such a way that each acquired pixel is converted into an electrical charge. CCDs are now commonly found in digital still and video cameras, telescopes, scanners and barcode readers. A camera uses an objective lens to focus the incoming light, and if we place a CCD array at the focused point where this optical image is formed, we can capture a likeness of this image (Figure 2.1).

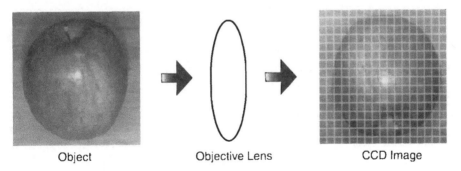

| Object | Objective Lens | CCD Image |

FIGURE 2.1 Camera schematic.

2.1.1 SCAN TYPES

The main types of cameras are broken into three distinct groups: (1) progressive area scan, (2) interlaced area scan and (3) line scan.

2.1.1.1 Progressive Area Scan

If your object is moving quickly, you should consider using a progressive scan camera. These cameras operate by transferring an entire captured frame from the image sensor, and as long as the image is acquired quickly enough, the motion will be frozen and the image will be a true representation of the object. Progressive-scan camera designs are gaining in popularity for computer-based applications, because they incorporate direct digital output and eliminate many time-consuming processing steps associated with interlacing.

2.1.1.2 Interlaced Area Scan

With the advent of television, techniques needed to be developed to minimize the amount of video data to be transmitted over the airwaves, while providing a satisfactory resulting picture. The standard interlaced technique is to transmit the picture in two pieces (or *fields*), called *2:1 Interlaced Scanning* (Figure 2.2). So an image of the letter "a" broken down into its 2:1 interlace scanning components would yield the result shown in Figure 2.3.

One artifact of interlaced scanning is that each of the fields are acquired at a slightly different time, but the human brain is able to easily combine the interlaced field images into a continuous motion sequence. Machine vision systems, on the other hand, can become easily confused by the two images, especially if there is excessive motion between them, as they will be sufficiently different from each other. If we consider the example in Figure 2.3, when the object has moved to the right between the acquisition of Field A and Field B, a resulting interlaced image would not be a true representation of the object (Figure 2.4).

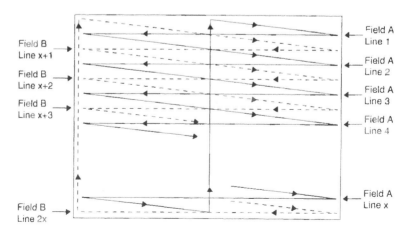

FIGURE 2.2 Interlaced scanning schematic.

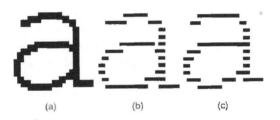

FIGURE 2.3 (a) Interlaced (b) Field A (c) Field B.

FIGURE 2.4 Interlaced image with object in motion.

2.1.1.3 Interlacing Standards

2.1.1.3.1 NTSC vs. PAL

The first widely implemented color television broadcast system was launched in the U.S. in 1953. This standard was penned by the National Television System Committee, and hence called *NTSC*. The NTSC standard calls for interlaced images to be broadcast with 525 lines per frame, and at a frequency of 29.97 frames per second. In the late 1950s, a new standard was born, called *PAL* (Phase Alternating Line), which was adopted by most European countries (except France) and other nations, including Australia. The PAL standard allow for better picture quality than NTSC, with an increased resolution of 625 lines per frame and frame rates of 25 frames per second.

Both of the NTSC and PAL standards also define their recorded tape playing speeds, which is slower in PAL (1.42 m/min) than in NTSC (2.00 m/min), resulting in a given length of video tape containing a longer time sequence for PAL. For example, a T-60 tape will indeed provide you with 60 min of recording time when using the NTSC standard, but you will get an extra 24 min if using it in PAL mode.

2.1.1.3.2 RS-170

RS-170 standard cameras are available in many flavors, including both monochrome and color varieties. The monochrome version transmits both image and timing information along one wire, one line at a time, and is encoded using analog variation. The timing information consists of horizontal synch signals at the end of every line, and vertical synch pulses at the end of each field.

The RS-170 standard specifies an image with 512 lines, of which the first 485 are displayable (any information outside of this limit is determined to be a "blanking" period), at a frequency of 30 interlaced frames per second. The horizontal resolution is dependent on the output of the camera — as it is an analog signal, the exact number of elements is not critical, although typical horizontal resolutions are in the order of 400 to 700 elements per line. There are three main versions of the color RS-170 standard, which use one (composite video), two (S-Video) or four (RGBS; Red, Green, Blue, Synchronization) wires. Similar to the monochrome standard, the composite video format contains intensity, color and timing information on the same line. S-Video has one coaxial pair of wires: one carries combined intensity and timing signals consistent with RS-170 monochrome, and the other carries a separate color signal, with a color resolution much higher than that of the composite format. S-Video is usually carried on a single bundled cable with 4-pin connectors on either end. The RGBS format splits the color signal into three components, each carrying high-resolution information, while the timing information is provided on a separate wire — the synch channel.

2.1.1.4 Line Scan

Virtually all scanners and some cameras use image sensors with the CCD pixels arranged in one row. Line scan cameras work just as their name suggests: a linear array of sensors scans the image (perhaps focused by a lens system) and builds the resulting digital image one row at a time (Figure 2.5).

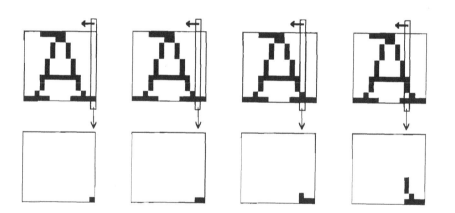

FIGURE 2.5 Line scan progression.

The resolution of a line scan system can be different for each of the two axes. The resolution of the axis along the linear array is determined by how many pixel sensors there are per spatial unit, but the resolution perpendicular to the array (i.e., the axis of motion) is dependent on how quickly it is physically scanning (or *stepping*). If the array is scanned slowly, more lines can be acquired per unit length, thus providing the capability to acquire a higher resolution image in the perpendicular axis. When building an image, line scan cameras are generally used only for still objects.

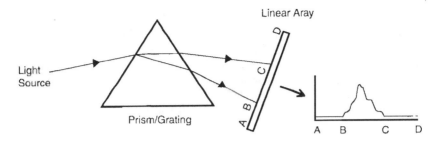

FIGURE 2.6 Spectrometer schematic.

A special member of the Line Scan family does not scan at all — the line detector. This type is often used when detecting intensity variations along one axis, for example, in a spectrometer (Figure 2.6).

As the light diffracts through the prism, the phenomenon measured is the relative angular dispersion, thus a two-dimensional mapping is not required (although several such detectors actually use an array a few pixels high, and the adjacent values are averaged to determine the output for their perpendicular position of the array).

2.1.1.5 Camera Link

A relatively new digital standard, Camera Link was developed by a standards committee of which National Instruments, Cognex and PULNiX are members. Camera Link was developed as a scientific and industrial camera standard whereby users could purchase cameras and frame grabbers from different vendors and use them together with standard cabling. Camera Link benefits include smaller cable sizes (28 bits of data are transmitted across five wire pairs, decreasing connector sizes, therefore theoretically decreasing camera-housing sizes) and much higher data transmission rates (theoretically up to 2.38 Gbps).

The Camera Link interface is based on Low Voltage Differential Signaling (LVDS), a high-speed, low-power, general purpose interface standard (ANSI/TIA/EIA-644). The actual communication uses differential signaling (which is less susceptible to noise — the standard allows for up to ± 1V in common mode noise to be present), with a nominal signal swing of 350 mV. This low signal swing decreases digital rise and fall times, thus increasing the theoretical throughput. LVDS uses current-mode drivers, which limit power consumption. There are currently three configurations of Camera Link interface — Base, Medium and Full — each with increasing port widths.

The Camera Link wiring standard contains several data lines, not only dedicated to image information, but also to camera control and serial communication (camera power is not supplied through the Camera Link cable).

More information regarding the Camera Link standard can be found in the "Specifications of the Camera Link Interface Standard for Digital Cameras and Frame Grabbers" document on the accompanying CD-ROM.

2.1.1.6 Thermal

Used primarily in defense, search and rescue and scientific applications, thermal cameras allow us to "see" heat. Thermal cameras detect temperature changes due to the absorption of incident heat radiation. Secondary effects that can be used to measure these temperature fluctuations include:

Type	Description
Thermoelectric	Two dissimilar metallic materials joined at two junctions generate a voltage between them that is proportional to the temperature difference. One junction is kept at a reference temperature, while the other is placed in the path of the incident radiation.
Thermoconductive	A *bolometer* is an instrument that measures radiant energy by correlating the radiation-induced change in electrical resistance of a blackened metal foil with the amount of radiation absorbed. Such bolometers have been commercially embedded into arrays, creating thermal image detectors that do not require internal cooling.
Pyroelectric	Pyroelectric materials are permanently electrically polarized, and changes in incident heat radiation levels alter the surface charges of these materials. Such detectors lose their pyroelectric behavior if subject to temperatures above a certain level (the Curie temperature).

Thermal cameras can be connected to National Instruments image acquisition hardware with the same configuration as visible spectrum cameras. More information regarding thermal cameras can be found at the *FLIR Systems* Web site (http://www.flir.com).

IndigoSystems also provides a large range of off-the-shelf and custom thermal cameras that ship with native LabVIEW drivers. More information on their products can be found at http://www.indigosystems.com.

2.1.2 CAMERA ADVISORS: WEB-BASED RESOURCES

Cameras come in all shapes, sizes and types; choosing one right for your application can often be overwhelming. National Instruments provides an excellent online service called "Camera Advisor" at http://www.ni.com (Figure 2.7).

Using this Web page, you can search and compare National Instruments tested and approved cameras by manufacturer, model, vendor, and specifications. There is also a link to a list of National Instruments tested and approved cameras from the "Camera Advisor" page. Although the performance of these cameras is almost guaranteed to function with your NI-IMAQ hardware, you certainly do not have to use one from the list. As long as the specifications and performance of your selected camera is compatible with your hardware, all should be well.

Another excellent resource for cameras and lenses system is the Graftek Web site (http://www.graftek.com; Figure 2.8). Graftek supplies cameras with NI-IMAQ drivers, so using their hardware with the LabVIEW Vision Toolkit is very simple. The Graftek Web site also allows comparisons between compatible products to be made easily, suggesting alternative hardware options to suit almost every vision system need

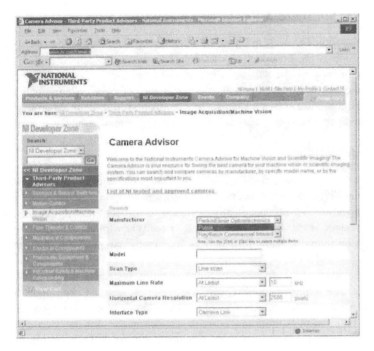

FIGURE 2.7 National Instruments Camera Advisor.

FIGURE 2.8 The Graftek website.

2.1.3 USER SOLUTION: X-RAY INSPECTION SYSTEM

Neal Pederson holds an M.S. in Electrical Engineering from Northwestern Polytechnic University, and is the president of VI Control Systems Ltd., based in Los Alamos. NM. His company, an NI Alliance Program Member, offers years of experience in software and hardware development of controls for complex systems such as linear accelerators and pulse-power systems, data acquisition in difficult environments and sophisticated data analysis. Neal can be contacted by e-mail at np@vicontrols.com.

2.1.3.1 Introduction

The 1-MeV x-ray inspection system will inspect pallets and air cargo containers in order to identify contraband (including drugs and weapons) without the need for manual inspection. While specifically designed to inspect air cargo pallets and air cargo containers, it can be used to inspect virtually any object that can fit through the inspection tunnel (Figure 2.9) such as machinery, cars and vans, and materials of all types.

The system combines two new technologies: the Nested High Voltage Generator (NHVG), a proprietary technology of North Star Research Corp., and MeVScan (U.S. Patent #6009146), a magnetically controlled moving x-ray source with a stationary collimator. It produces x-ray pencil beams for transmission and scatter imaging. Two transmission images are produced, 30° apart. This gives the operator a stereoscopic, three-dimensional view of the inspected cargo and increases the likelihood of seeing objects hidden behind other objects. Additionally, a backscatter image is produced. These 3 images are processed and displayed separately at the operator's control console.

FIGURE 2.9 X-Ray inspection system.

2.1.3.2 Control and Imaging Hardware Configuration

The remote computer control system is mounted in a bay of four 24-in. wide racks. There are four 21-in. monitors controlled by one computer and a multi-monitor video card. There is also one CRT with four video sources that is used for video surveillance of the system. One of the four computer monitors is used for controlling the scan process and for entering user and pallet information. The other three monitors are used to display and manipulate the three x-ray images (Transmission 1, Transmission 2 and Backscatter). The images are generated by acquiring PMT data on a PCI-6052E DAQ board. One thousand vertical pixels are acquired for each x-ray scan. This is duplicated 1000 times in the horizontal direction while pallets are moving on the conveyors to create a 1000 × 1000 pixel image for each of the three images. A PC-TIO-10 timing board carefully times the x-ray system and the data collection. The Remote Device Access Server (RDA) that is built into NI-DAQ and a fiber-optic Ethernet link allow the data to be acquired from a remote computer. This isolates the operator's console from high voltage and prevents EMI/RFI problems. RDA provides the full 333 kS/s bandwidth of the PCI-6052E DAQ board in a double-buffered mode. All DC analog and digital controls use Group3's fiber-optically distributed control hardware. The Group3 hardware is specifically designed for accelerator and high voltage applications where isolation is critical.

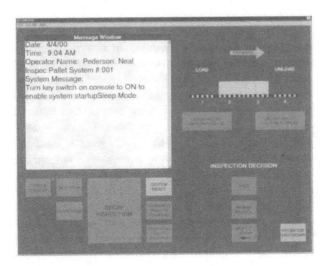

FIGURE 2.10 Control screen.

2.1.3.3 Control Software

The control software (Figure 2.10) was designed to allow operators with little or no computer experience to inspect pallets without having any knowledge of how the x-ray machine works. This meant that a very complicated process of control, conditioning and fault detection had to be totally automated. This was accomplished by using a series of LabVIEW state machines. Four state machines were used to handle the major control processes:

1. *Cold start procedure:* initial log-on and vacuum system conditioning
2. *Warm start procedure:* repeat log-ons and NHVG preparation
3. *Scan procedure:* conveyor automation, x-ray generation and image data collection
4. *Fault/reset procedure:* fault handling and system reset

The automatic state machines control approximately 200 I/O points. A maintenance mode will bypass the automatic state machines and allow maintenance personnel to control any I/O point individually. This was very useful during system development and troubleshooting.

A system administrator screen allows the system administrator to provide names, passwords and access levels to users. User names are included in the header of each archived image and passwords prevent regular users from going into the maintenance and administrator modes.

2.1.3.4 Imaging Software

The imaging software handles display and manipulation of x-ray image data. The imaging functions run on separate screens and take advantage of the multi-threaded parallel loop capabilities that are built into LabVIEW. There are three screens dedicated to image manipulation; each screen has an x-ray image, a thumbnail copy of the x-ray image and a control palette for controlling image functions (Figure 2.11).

When image data is acquired, the data is first processed and then buffered in 16 bit integer format. A fourth x-ray detector is used to normalize the data while it is being collected. Because the x-ray detectors pick up cosmic rays from outer space, a software routine is used to eliminate them before display. By double-buffering the data while it is being collected, the Transmission 1 image is updated 20 scans at a time while the images are being acquired. Transmission 2 and Backscatter images are updated after the scan is complete. The 16 bit data is converted to Unsigned 8 bit integer format before it is displayed on the screen as 256 grayscale.

Data for the x-rays has much higher resolution than can be seen with 256 grayscale. A process called *window and level* is used to "see into" the x-ray data. The window and level routine allows the operator to set the upper and lower depths of x-ray penetration that is displayed for each image. This has the affect of being able to see through materials of different densities.

IMAQ Vision is used to display and manipulate the image data. These functions include zoom, pan, reverse video, edge enhance and 3-D position, which is used to find the vertical height of an object after the operator makes a selection on each of the two stereoscopic transmission images. A thumbnail of each image displays the entire image in a 250×250 pixel area. If the image is zoomed beyond $1\times$, the operator can simply click on the portion of interest in the thumbnail image and the zoomed image will pan to this portion of the image.

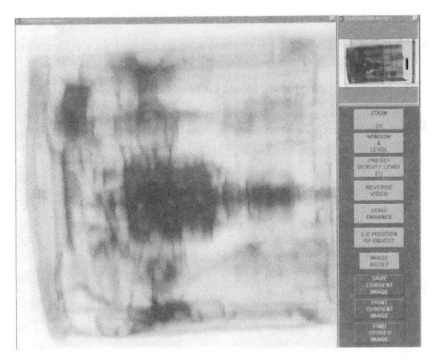

FIGURE 2.11 X-Ray imaging screen (front half of a van).

2.1.3.5 Conclusions

The combination of LabVIEW, IMAQ Vision and NI-DAQ RDA provided a complete and economical virtual instrumentation solution to the control of a very complex x-ray machine. Other software and hardware packages could have done each piece of this system, but it would have been difficult and expensive to integrate. LabVIEW provided fast software development that could leverage off other Windows technologies such as multi-monitor video cards and TCP/IP communications. IMAQ Vision provided powerful imaging tools that easily combined the data acquisition and control capabilities of LabVIEW. NI-DAQ RDA provided a seamless integration between high-speed DAQ boards and a remotely isolated control location.

2.2 IMAGE ACQUISITION HARDWARE

Image acquisition hardware is the physical interface between your PC and camera, such as frame grabbers and embedded interfaces that come standard with most PCs.

2.2.1 NATIONAL INSTRUMENTS FRAME GRABBERS

National Instruments image acquisition hardware (Figure 2.12) is often the first choice for developers using LabVIEW and the Vision Toolkit. It works seamlessly with the software, and is specifically designed for integration with other National Instruments products such as DAQ, motion control and CAN hardware.

FIGURE 2.12 A populated PXI-1020.

National Instruments has a small range of image acquisition cards for both PCI and PXI (Figure 2.13) architectures, and they all have either the RTSI or PXI trigger buses, which makes synchronizing IMAQ and DAQ very simple and accurate. Although its IMAQ hardware is often considered expensive, the quality and functionality of the cards often outweighs the excessive cost of subsequent development time when a card from a different vendor is chosen. Comprehensive information on the National Instruments image acquisition hardware range can be found at http://www.ni.com.

2.2.2 IEEE 1394 (FIREWIRE) SYSTEMS

First conceived in the distant past (as far as PC technology is concerned), the IEEE 1394 standard enables simple, high-bandwidth data interfacing between computers, peripherals and consumer electronics products.

(a)   (b)

FIGURE 2.13   (a) National Instruments Frame Grabber (NI-1409). (b) IMAQ cards in a PXI rack.

Apple Computer technologists (www.apple.com) defined the standard in 1986, naming it *FireWire* in reference to the high data speeds that it could sustain — far quicker than any other technology of its type at the time. Adopted by the IEEE in 1995, FireWire has become extremely popular as PC and operating system vendors include embedded support for it. For example, both Dell Computers (www.dell.com) and Microsoft® Windows® (www.microsoft.com/windows) now include native support for FireWire hardware, and National Instruments retails a Measurement and Automation Explorer plug-in that extends its capability to interfacing with FireWire cameras.

The high data rates of FireWire, the ability to mix real-time and asynchronous data on a single connection, and simultaneous low-speed and high-speed devices on the same network are all very strong reasons to use FireWire hardware. The FireWire bus can have up to 63 devices simultaneously connected, although IEEE 1394.1 bus bridges will soon be available, extending the capability to more than 60,000 devices. There are currently three standard IEEE 1394 signaling rates: S100 (98.304 Mbps), S200 (196.608 Mbps) and S400 (393.216 Mbps), although two new standards that should be available very soon will reach speeds of approximately 800 and 1200 Mbps. Thankfully, these faster busses are designed to support existing lower speed devices as well, and even mixed speeds simultaneously.

(a)   (b)

FIGURE 2.14 (a) PCI IEEE 1394 card. (b) Integrated laptop IEEE 1394 port.

Getting data into your PC requires an IEEE 1394 adapter, which can either be a PCI card, or an integrated interface (Figure 2.14). There are two accepted plug-socket standards: six-pin (shown on the PCI card in Figure 2.14(a)) and four-pin (Figure 2.14(b)). The IEEE 1394 standard defines that the six-pin systems contain two separately shielded twisted pairs for data transmission, plus two power conductors and an overall external shield. These power conductors are used to supply power to devices on the bus (between 8 and 30 volts DC, with a maximum loading of 1.5 A), whereas the smaller four-pin conductors are for data transmission only. This is particularly important to remember when attempting to connect devices that draw their power from the IEEE 1394 bus, as they will not work when using the four-pin variety.

FireWire cables are limited to 4.5 m between devices, although restricting the speed of the bus to the S200 standard allows a distance increase to approximately

14 m. The use of IEEE 1394 repeaters (Figure 2.15) has become common when the full S800 speed is required, and IEEE 1394 transceivers have been announced that are powered by in-wall wiring and extend the distance between active nodes to at least 70 m using plastic optical fiber.

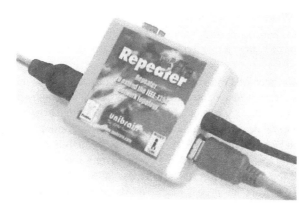

FIGURE 2.15 IEEE 1394 repeater.

As noted previously, these quoted distances are between each device, so a standard network containing many FireWire devices can be up to 63 devices × 4.5 m = 283.5 m.

There are a large number of FireWire cameras currently available (Figure 2.16), including models from Basler Vision Technologies (www.baslerweb.com), Sony (www.sony.com), unibrain (www.unibrain.com) and SVS-VISTEK GmbH (www.svs-vistek.com).

FIGURE 2.16 IEEE 1394 camera with lens.

More information regarding the IEEE 1394 standard and applicable hardware can be found at the IEEE 1394 Trade Association Web site (http://www.1394ta.org).

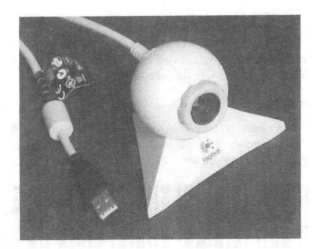

FIGURE 2.17 An inexpensive webcam with a USB interface.

2.2.3 USB SYSTEMS

Although very inexpensive, *universal serial bus* (USB) cameras often suffice when your image acquisition needs are simple (Figure 2.17). USB cameras are small, easily connected to your PC, and do not require a frame grabber card.

There are currently two USB standards: the original version is able to handle data rates of 1.5 Mbps, whereas the new version (USB 2.0) can transmit at 12 Mbps. USB 2.0 can theoretically reach transfer rates of 480 Mbps; however, there is a wide variation in edge rates. With typical line loads, full speed devices usually fall in the 12 to 25 ns range, and low-speed devices typically range 110 to 225 ns.

The USB standard allows for a limit of 127 devices to be connected simultaneously, although devices often reserve large ranges of the available bandwidth, so this limit is rarely practically achievable. If a large number of devices are required, you can install extra USB interfaces in your PC, effectively providing independent USB buses and raising the number of devices that will function simultaneously. USB cable lengths are limited to 5 m due to a cable delay specification of 26 ns that allows for signal reflections to settle in the cable before any subsequent bits are transmitted.

Both USB and IEEE 1394 are serial buses; IEEE 1394 can move more data in a given amount of time, but is considerably more expensive than USB. Applications that are best suited for FireWire are high-bandwidth applications (e.g., external disk drives and high-quality video streams), whereas USB fills the middle- and low-bandwidth application niche (including Webcams, audio, scanners and printers).

National Instruments does not provide native support for USB cameras, although several LabVIEW drivers are available. One such solution is Peter Parente's *Webcam Library*, which is featured in Chapter 3.

More information regarding the USB standards and applicable hardware can be found at the USB Implementers Forum Web site (http://www.usb.org).

2.3 FROM OBJECT TO CAMERA

Often the physical constraints of the object you are monitoring or existing hardware around it will control the placement of your camera, lens system and lighting. Before you buy any hardware, you should consider several factors.

2.3.1 RESOLUTION

The resolution of a camera system refers to the smallest feature of the object that it can distinguish. A common example used to demonstrate resolution is a barcode, shown in Figure 2.18. As you can see, a barcode is made from black and white bars

FIGURE 2.18 Barcode.

of varying width. If the resolution of your acquisition system is too coarse to resolve between the smallest black and white bars, then useful data will be lost, and you will not be able to resolve the barcode. One common solution is to use the highest resolution system available, and then you will never lose useful data. However, there are several reasons not to choose this option: it requires more processing power and memory, is much slower and is much more expensive. The tricky part is to determine a suitable compromise, and therefore most efficient and economical resolution.

As a general rule, your smallest useful object feature should be at least 2 pixels. If so, you can be assured of never missing it due to its image falling on an inactive part (sections between each of the CCD array elements) of your detector. Although that sounds simple, you need to consider the lens system that you are using to focus your image on the detector array. For example, using a 2× zoom system will effectively increase the resolution of your image, but decrease its field of vision. To determine the sensor resolution, you will require to achieve a smallest feature of 2 pixels, you can use the following equation:

$$\text{Sensor Resolution}_{1D} = \left\lceil 2\left(\frac{FieldOfView}{SmallestFeature} \right) \right\rceil$$

The sensor resolution is a dimensionless quantity; use the same units for both the field of view and smallest feature to achieve a useful answer. For example, when reading a barcode 35 mm long, with a narrow bar of 0.75 mm, through a lens system of 1× magnification, the sensor resolution would need to be at least:

$$\text{Sensor Resolution}_{1D} = \left\lceil 2\left(\frac{35}{0.75} \right) \right\rceil = 94 \, pixels$$

Although this would be useful for a barcode (as the information you are attempting to acquire is basically a one-dimensional line profile across the code), most detector arrays are two-dimensional, so this equation needs to be applied across both dimensions. Another example is measuring the size of an apple. Consider a particular variety of apple, which has a maximum horizontal size of 8 cm, its vertical size will not exceed 10 cm, and the required accuracy of the measurement needs to be 1 mm (0.001 m). The minimum sensor resolution is determined as:

$$\text{Sensor Resolution}_H = \left[2 \left(\frac{0.08m}{0.001m} \right) \right] = 160\,pixels$$

$$\text{Sensor Resolution}_v = \left[2 \left(\frac{0.10m}{0.001m} \right) \right] = 200\,pixels$$

indicating a sensor with a resolution of at least 160 × 200 is required.

2.3.2 DEPTH OF FIELD (DOF)

The DOF of a lens system is the lens-perpendicular spatial range of which objects remain in focus; any objects outside of the DOF will not be in focus. The DOF is directly related to the *blur diameter*, as shown in Figure 2.19. Consider a beam (a group of adjacent rays) of light traveling from the object through the lens and focusing on the CCD array of a camera. As the beam travels through the lens, it undergoes geometric aberration, both spatially and temporally, so the beam defocuses. If the beam is defocused enough to cover an area of the CCD that is larger than the required resolution (the blur spot), then the image will be out of focus. This phenomenon also can occur due to chromatic aberration, which is evident when different frequencies of light (e.g., colors in the visible spectrum) are diffracted at different angles.

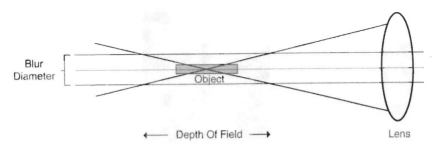

FIGURE 2.19 Depth of field.

One method of increasing DOF is to place a limiting iris in front of the lens (Figure 2.20). Unfortunately, using this method causes less light to irradiate the lens, so you will need to increase the amount of light falling on the object to achieve similar contrast levels.

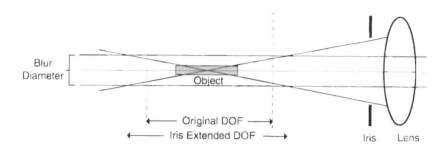

FIGURE 2.20 Decreased iris.

2.3.3 CONTRAST (OR MODULATION)

Contrast is often referred to as "resolution's brother." Resolution is the ability to resolve spatial differences in your image, while contrast is the ability to resolve intensity differences. Contrast is therefore a dimensionless quantity that is the difference between the lightest and darkest features of an image:

$$Contrast = \frac{I_{Brightest} - I_{Darkest}}{I_{Brightest} + I_{Darkest}} \; where \; I_{Brightest} > I_{Darkest}, I_{Darkest} \neq 0$$

As this equation suggests, when the difference between $I_{Brightest}$ and $I_{Darkest}$ is high, the contrast tends toward 1 (unity), indicating that the image has a large range of contrast. Conversely, when the image is "flat" and has very little contrast, the contrast value approaches 0. In an example image

123	122	117	125	134
132	120	115	122	133
131	121	118	125	132
133	125	119	127	130
132	129	120	122	129

$$I_{Brightest} = 134, \; I_{Darkest} = 115$$

$$\therefore Contrast = \frac{134 - 115}{134 + 115} = 0.076$$

indicates a low-contrast image, whereas the following example image:

123	180	193	221	247
186	183	149	173	185
255	208	89	143	136
250	210	11	78	102
246	205	15	101	66

$$I_{Brightest} = 255, \quad I_{Darkest} = 11$$

$$\therefore Contrast = \frac{255 - 11}{255 + 11} = 0.917$$

indicates an image with a high-contrast level. A high-contrast image often appears to be sharper than that of a lower contrast image, even when the acquired image is of identical resolution.

2.3.4 PERSPECTIVE (PARALLAX)

Perspective errors occur when the object's surface is not perpendicular to the camera axis (Figure 2.21). The second image is quite badly distorted (skewed) due to the high angle between the object's surface and the camera axis.

Perspective errors occur in just about every lens system that exists, unless the only object feature that you are interested in is directly below the camera axis, and is only a few pixels wide. There are two common methods of minimizing perspective error: software calibration and telecentricity.

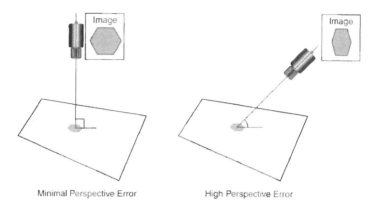

Minimal Perspective Error   High Perspective Error

FIGURE 2.21 Camera perspectives.

2.3.4.1 Software Calibration

Calibration occurs when an object with known features is acquired, and the differences between the known object and the acquired image are used to process subsequently acquired images. The Vision Toolkit ships with several intuitive calibration routines and known grids to help make calibration easier. Consider the standard Vision Toolkit Calibration Grid in Fi gure 2.22. This grid is provided

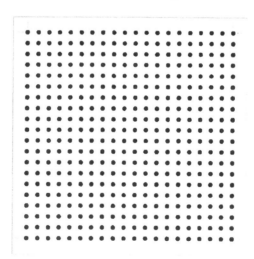

FIGURE 2.22 Calibration grid.

with the Vision Toolkit as a PDF document, and can found using the following path: Start > Programs > National Instruments > Vision > Documentation > Calibration Grid.

The calibration grid's dots have radii of 2 mm and center-to-center distances of 1 cm, although this may vary slightly depending on your printer setup (using a high-resolution printer will improve the accuracy of a calibration). National Instruments suggests that once you have printed it out:

- The displacement in the x and y directions be equal.
- The dots cover the entire desired working area.
- The radius of the dots is 6 to 10 pixels.
- The center-to-center distance between dots range from 18 to 32 pixels.
- The minimum distance between the edges of the dots is 6 pixels.

If you cannot print the grid to the specifications recommended by National Instruments, you can purchase preprinted grids from several companies such as Edmund Industrial Optics (go to http://www.edmundoptics.com and type "calibration target" in the Search box). If we acquire an image of a calibration grid using the same camera setup as described previously, the perspective error is quite apparent (Figure 2.23).

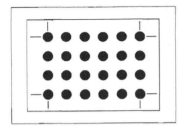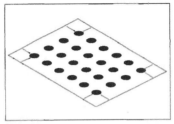

FIGURE 2.23 Calibration grid with perspective error.

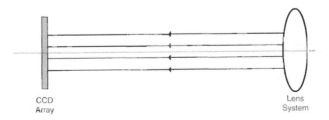

FIGURE 2.24 Telecentricity.

2.3.4.2 Telecentricity

Telecentricity occurs in dedicated multi-element lens systems when all of the chief rays for all points across the object are collimated (parallel to the optical axis) (Figure 2.24). Telecentricity can be difficult to visualize, as it provides nearly constant magnification over a range of working distances, virtually eliminating perspective angle error across the face of the lens.

Consider a car approaching you. As it gets closer, it looks larger, but if it is viewed with a telecentric lens, its size will not change while it is in the lens' telecentric range. Another advantage of using a telecentric lens is the parallel nature of the chief rays: several objects' endpoints can be easily inspected without their three-dimensional nature distorting the image.

Consider a camera acquiring an object that is not directly under it (Figure 2.25). A traditional lens system forms an image with the top of the box distorted, and the sidewalls visible (also distorted), whereas due to the parallel nature of the chief rays, the telecentric lens forms the undistorted image of only the top surface (Figure 2.26). There are several disadvantages to using a telecentric lens, including the optics sizes. When using a doubly telecentric design (which refers to a lens system that is telecentric in both object and image space), both the front- and rear-most lens groups must be larger than the object and image, respectively. If you have an object that is 15 cm wide, then a much larger aperture diameter is required to provide a clean image (free of any retention and mounting hardware spoiling the image).

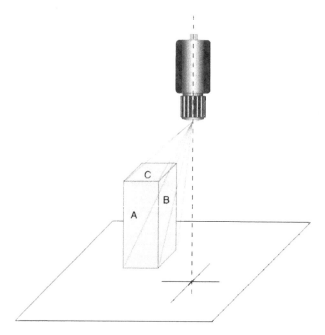

FIGURE 2.25 Object off camera axis.

A common misconception is that telecentric lenses have very large DOFs: this is not necessarily true (unless the lens was specifically designed to have a large DOF), as a lens system's DOF depends only on its focal length (f-number) and resolution. With telecentric lenses, objects still blur when they are outside of the DOF, although any blurring occurs symmetrically across the image.

Further telecentricity information can be found at the Edmund Industrial Optics Web site (http://www.edmundoptics.com/TechSupport/DisplayArticle.cfm?articleid = 239).

An excellent manufacturer of telecentric lenses is the Liebmann Optical Company (http://www.jenoptik.com; a part of JENOPTIK AG, which grew out of JENOPTIK Carl Zeiss JENA GmbH in 1991). Further product information (as well as

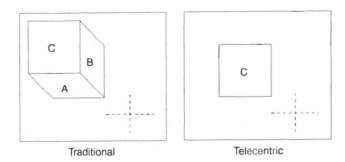

Traditional   Telecentric

FIGURE 2.26 Traditional vs. telecentric lens systems.

several other brands of lenses and machine vision components) can be found at the Adept Electronic Solutions Web site: (http://www.adept.net.au).

2.3.5 NATIONAL INSTRUMENTS LENS PARTNERS

National Instruments has several lens partners that have been selected due to their wide range of quality lens systems designed to assist you in completing a vision system. These partners include Edmund Industrial Optics (http://www.edmundoptics.com), Navitar (http://www.navitar.com), Infinity Photo-Optical (http://www.infinity-usa.com) and Graftek Imaging (http://www.graftek.com). Graftek Imaging's Web site has a "lens selector." You can specify lens parameters, and it will not only find all possible matches within their product range, but also display possible compatible products for comparison (Figure 2.27).

FIGURE 2.27 The Graftek lens selector Website.

2.4 LIGHTING

Lighting is an important, yet often overlooked facet of image acquisition — lighting can make or break your system. Some system integrators have been known to consider the industrial lighting solution a low-priority task, to be quickly decided upon toward the very end of the project during the implementation phase. Although this approach can be fruitful, selecting an appropriate lighting apparatus during the project definition stage can drastically change the path of your development process. Selecting a different illumination solution, rather than investing in higher resolution detectors and imaging lenses, can improve often image quality. Although machine vision lighting hardware can come in many shapes and sizes, the most common types are area array, backlight, dark field and ring and dome illuminators. Specialized types include cloudy day, diffuse on axis, collimated on axis and square continuous diffuse illuminators.

2.4.1 AREA ARRAYS

Area array illuminators are by far the most common machine vision lighting devices, and are typically used on nonspecular surface objects with low reflection. The physical makeup of an area array is a matrix of small lights (often LEDs) covering a large area, thus creating a wide beam of light toward the object. The most common area array variants include spot and flat area array sources (as shown in Figure 2.28). Area array sources are often the cheapest solution to general purpose light applications, and are especially useful when the light source is required to be some distance from the object (spot lighting).

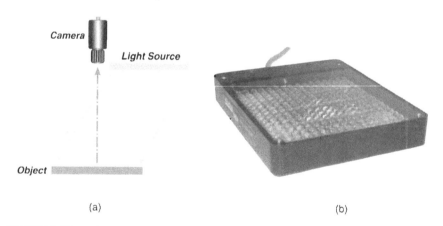

| (a) | (b) |

FIGURE 2.28 (a) Area array schematic. (b) Area array.

2.4.2 RING

Ring illumination is a popular choice for applications that require coaxial lighting, and can produce uniform illumination when used at appropriate distances. As the ring can be placed around the viewing axis, the object is irradiated almost as though the light is projecting from or behind the camera itself (Figure 2.29). When used in this method, the resulting image generally has few shadows. Ring lighting is only useful when illuminating nonreflective surfaces, otherwise circular glare patterns are reflected (Figure 2.30).

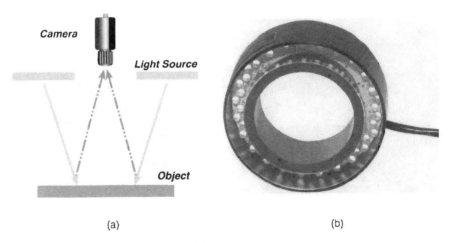

(a)   (b)

FIGURE 2.29 (a) Ring schematic. (b) Ring

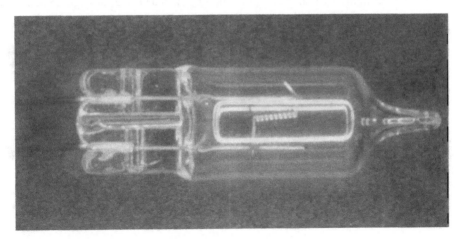

FIGURE 2.30 Reflective object illuminated with a ring light.

Ring lighting is often used in areas such as microscopy, photographic documentation, medical and biological research, metallurgical and forensic inspection, semiconductor etching and printing inspection. Ring lights are particularly useful when illuminating objects at short working distances, otherwise its effects become similar to a point source or area array.

2.4.3 DARK FIELD

The light emitted from dark field illuminators (Figure 2.31) provides low-angle illumination, enhancing the contrast of surface features, and displaying the contours of the object, including engraved marks and surface defects.

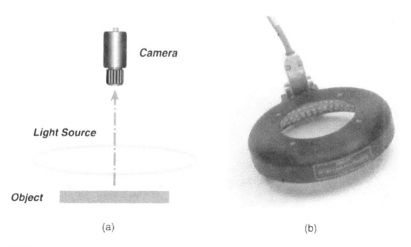

(a)	(b)

FIGURE 2.31 (a) Dark Field schematic; (b) Dark field.

Dark field lights cast shadows from the object, much like an edge detection image processing routine will enhance the perceived edges (where there is a large variation in intensity) in an image. The physical layout of the illuminator is similar to ring lighting, except that the illuminator is placed closer to the object, and the light is angled inward (Figure 2.32).

(a)	(b)

FIGURE 2.32 (a) Semireflective object illuminated with a ring light. (b) The same object illuminated with a dark field light.

2.4.4 DOME

Dome illuminators provide diffused uniform light, evenly illuminating three-dimensional surfaces, and therefore produce very few shadows. The dome light source is relatively close to the object under inspection. Domes lack 100% on-axis capability, leaving the light/camera aperture visible in the center of resulting images. If that area is not part of the region of interest, domes can offer a cost-effective alternative to continuous diffuse illuminators (Figure 2.33).

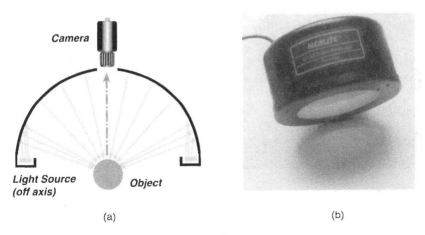

(a)   (b)

FIGURE 2.33 (a) Dome schematic. (b) Dome field.

2.4.5 BACKLIGHTS

Backlights are useful in providing silhouettes and enhanced images of internal components of clear or semiclear objects (Figure 2.34).

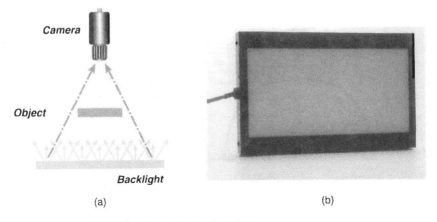

(a)   (b)

FIGURE 2.34 (a) Backlight schematic. (b) Backlight.

   In Figure 2.35, a lamp globe filament is difficult to resolve using a ring light for illumination, as excess light pollution is reflected from the glass surface. This reflected light is eradicated when the globe is backlit with a diffuse source, allowing much easier filament recognition. Backlights can also be useful in detecting cavities in semitransparent objects, and surface-to-surface breaches in opaque objects (e.g., drilled holes). Image processing can then be useful to count, map and classify these artifacts.

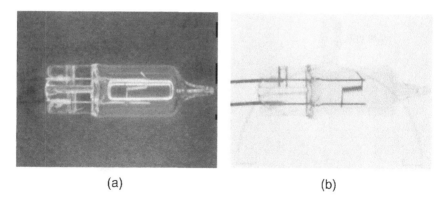

(a)   (b)

FIGURE 2.35 (a) Reflective object illuminated with a ring light. (b) The same object illuminated with a backlight.

2.4.6 CONTINUOUS DIFFUSE ILLUMINATOR (CDI™)

CDI creates almost infinitely diffuse light, and is particularly useful when inspecting highly specular surfaces where any reflections may cause a vision system to see defects generated by reflected light pollution (Figure 2.36).

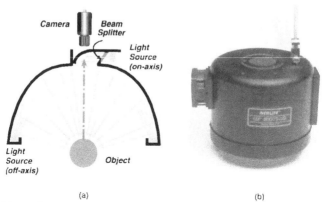

(a)   (b)

FIGURE 2.36 (a) Cloudy day schematic. (b) Cloudy day illuminator.

CDIs emulate cloudy days where only nonparallel diffuse light rays emanate from the sky. CDI applications include packaged component inspection, e.g., pharmaceutical blister-packages and electronic component tubes. The printed foil example in Figure 2.37 shows the reflected light when the object is illuminated with a ring light; it makes pattern recognition and therefore character recognition very difficult, whereas the diffuse multi-angled light decreases the glare from light pollution dramatically when using a CDI.

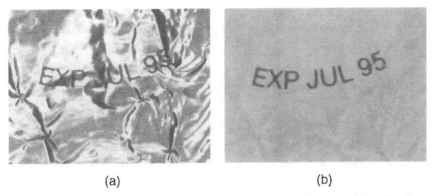

(a)   (b)

FIGURE 2.37 (a) An object illuminated with a ring light. (b) The same object illuminated with a CDI.

2.4.7 DIFFUSE ON AXIS LIGHTS (DOAL™) AND COLLIMATED ON AXIS LIGHTS (COAL)

DOAL and COAL lights use beam splitter technology to illuminate the object as though the camera is on the same axis as the light source (Figure 2.38).

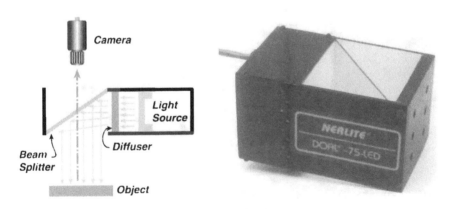

FIGURE 2.38 (a) DOAL schematic. (b) DOAL illuminator.

If desired, the light source can seem to be virtually at the same point as the camera. Specular surfaces perpendicular to the camera appear illuminated, while surfaces at an angle to the camera appear darker. Nonspecular surfaces absorb the incident light and appear dark. DOAL illuminators emit diffuse light, whereas COAL illuminators are designed for applications requiring coaxial bright light. The electronic circuit board example in Figure2.39 shows that the contrast achieved from the DOAL setup is much higher than if a ring light was used. This makes pattern recognition and caliper measurement much more reliable in an automated vision application.

Unfortunately, DOALs often have a low reflected light throughput to the camera, such that you may need to invest in multiple sources to provide sufficient detector irradiation.

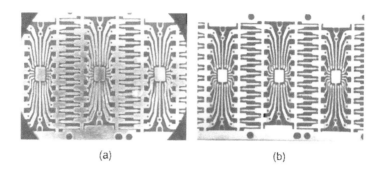

(a)   (b)

FIGURE 2.39 (a) PCB tracks illuminated with a ring light. (b) Illuminated with a DOAL.

2.4.8 SQUARE CONTINUOUS DIFFUSE ILLUMINATORS (SCDI™)

SCDI illuminators take the DOAL (described earlier) design to another level, providing significantly greater uniformity of diffuse illumination (Figure 2.40).

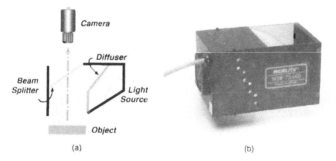

(a)   (b)

FIGURE 2.40 (a) SCDI schematic. (b) SCDI illuminator.

SCDIs are particularly suited to moderately faceted and undulating specular surfaces. Common examples of SCDI applications include compact disk artwork verification and the inspection of solder patterns on circuit boards (Figure 2.41).

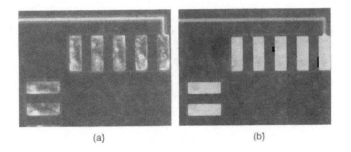

FIGURE 2.41 (a) PCB pads illuminated using a ring light. (b) Illuminated with a SCDI.

3 Image Acquisition

There are two main methods of obtaining images: loading from disk and acquiring directly using a camera. Image acquisition is often considered a complex task, but it can be simplified for most vision applications. National Instrument's Measurement and Automation Explorer (MAX), LabVIEW and the Vision Toolkit make image acquisition a straightforward and integrated task.

3.1 CONFIGURING YOUR CAMERA

The range of cameras available to use with your vision system is astounding – several different types exist across brand lines (more information of camera types can be found in Chapter 2). If you are using a National Instruments image acquisition card, MAX depends on a camera definition file to understand the configuration of the camera attached to the card. Configuration files for common cameras are included with a standard NI-IMAQ installation (Figure 3.1)

An icd file contains all of the information MAX and subsequently IMAQ needs to acquire images and control the camera. Figure 3.2 illustrates an example icd file. As shown, the icd file is of a human-understandable format, and can be manually edited using Notepad, or a similar ASCII text editor. As the camera configuration file extension is icd, the Windows® operating system should not have an application registered to launch the file type by default; therefore, you will need to manually open the file for

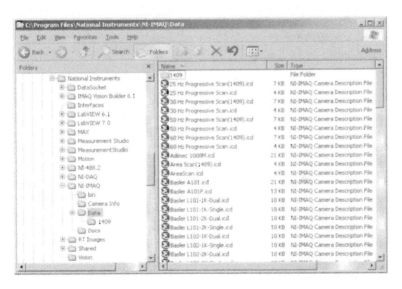

FIGURE 3.1 icd file directory.

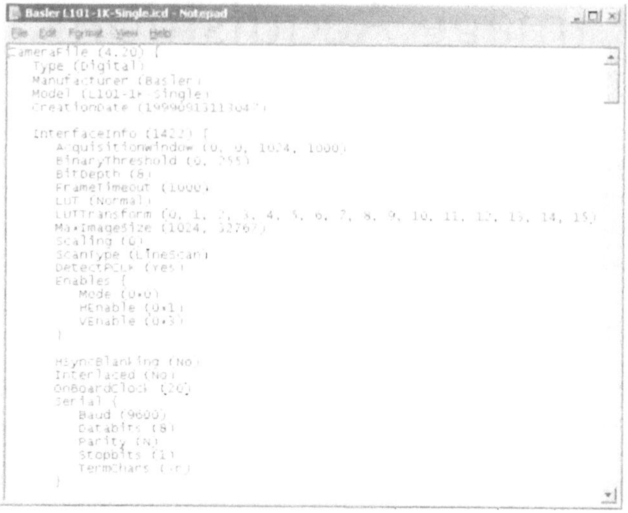

FIGURE 3.2 A example icd file.

viewing and editing. These files reside in the \National Instruments\NI-IMAQ\Data\ directory on your hard disk.

Be sure to select the icd file that matches both your camera and frame grabber. The icd file's name often describes the appropriate interface to use with the file; for example, camera files will often have an interface suffix denoting the interface type:

```
Camera_Name.icd
Camera_Name(1409).icd
Camera_Name(BNC).icd
```

If your camera does not have a corresponding icd file listed in the directory above, camera vendors will often provide one for you, or it may be available from National Instruments. The National Instruments FTP site stores icd files for common cameras, and can be accessed at ftp://ftp.ni.com/support/imaq/camera_support (Figure 3.3).

Fortunately, if you cannot obtain an appropriate file from the camera vendor, you can either create one, or copy and modify a file with a similar configuration to your camera.

If you are using a IEEE 1394 (FireWire) camera, an icd file is not required. NI-IMAQ will create one when the camera is detected. The created files are placed in the \National Instruments\NI-IMAQ for IEEE 1394\Data directory. Once you have a camera interface installed, you can set its properties using MAX (Figure 3.4).

Under Devices and Interfaces, right click on the interface (or camera name in the case of a FireWire camera), and select properties (Figure 3.5).

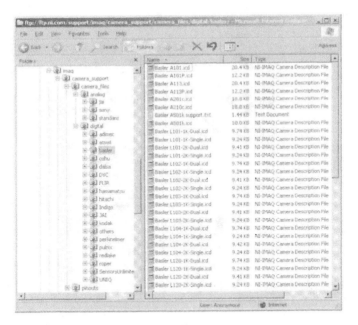

FIGURE 3.3 National Instruments icd file FTP site.

FIGURE 3.4 MAX camera configuration.

Depending on the hardware installed and the capabilities of your camera, you can set configuration parameters such as *brightness, autoexposure, sharpness, white balance, saturation, gamma, shutter speed, gain* and *iris*.

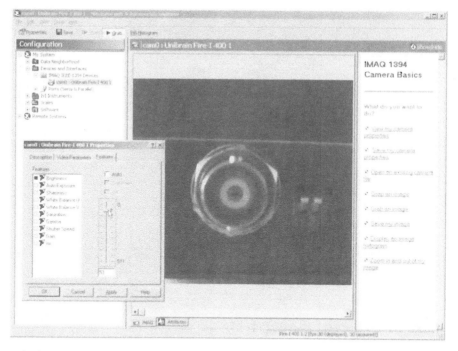

FIGURE 3.5 MAX camera properties.

FIGURE 3.6 IMAQ session properties.

3.2 ACQUISITION TYPES

There are four image acquisition types supported in the Vision Toolkit: Snap, Grab, Sequence and StillColor. Although the type your application needs may at first seem obvious, taking the time to understand the different types could save unnecessary development time and computation during postacquisition processing.

NI-IMAQ image acquisition often uses information regarding the open IMAQ session. To determine a plethora of camera, image and session properties, you can use a property node[1] (Figure 3.6). The property node example in Figure 3.8 returns the *image type* of the session, which is determined by either camera initialization, or is set dynamically. This *image type* value can then be passed to IMAQ Create, reserving the correct number of bytes per pixel for the acquired image data.

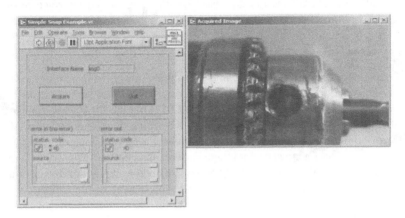

FIGURE 3.7 Simple snap example.

3.2.1 SNAP

Acquiring an image using IMAQ Snap is the simplest type of image acquisition (Figure 3.7). Used for slow acquisition rates, IMAQ Snap requires just three steps to execute: initialize, acquire and close. To initialize the IMAQ session, the *interface name* (or the *camera name* in the case of an FireWire interface) is input to IMAQ Init, which opens the camera by loading and parsing its icd file from disk (as determined by the Measurement and Automation Explorer), and opening an IMAQ session (Figure 3.8).

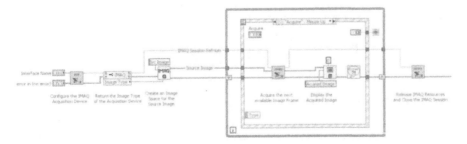

FIGURE 3.8 Simple snap example — wiring diagram.

The IMAQ session refnum is then fed into a property node, which is set to determine the image type of the camera interface. The dataspace for the image is

---

[1] The FireWire image acquisition VIs do not use a property node for setting acquisition properties. Use IMAQ1394 Attribute to set properties, and IMAQ1394 Attribute Inquiry to read properties.

created, and the system then waits for the user to click either *Acquire* or *Quit*. If the user clicks *Acquire*, IMAQ Snap is called (if IMAQ Init has not been previously called, the image acquisition board is now initialized, increasing the delay before acquisition), and the next available video frame from the camera is returned as *Image Out*. Finally, the IMAQ session is closed and the camera and interface resources released by calling IMAQ Close.

3.2.2 GRAB

Depending on the frame rate of the camera, grabbing images from an interface buffer is a very fast method of image acquisition. Using IMAQ Grab is the best method of acquiring and displaying live images (Figure 3.9). After the IMAQ session has been initialized, IMAQ Grab Setup is executed. Describing this VI as a "setup" VI is a slight misnomer, as it actually starts the camera acquiring images. IMAQ Grab Acquire is used to grab the current image in the camera buffer. IMAQ Grab Acquire has a Boolean input that defines when the next image is to be acquired: synchronized on the next vertical blank (i.e., wait for the next full image), or immediately transfer the current image (or part thereof) from the buffer.

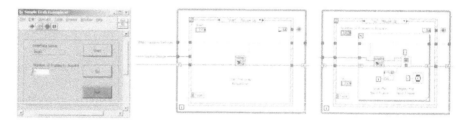

FIGURE 3.9 Simple grab acquisition – wiring diagram.

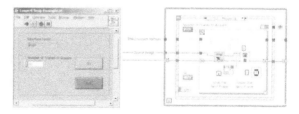

FIGURE 3.10 Looped Snap example.

A series of images can also be acquired using IMAQ Snap, but as it initializes the image interface every time it is called, the acquisition rate can be very slow (in the order of one frame per second) (Figure 3.10).

3.2.3 SEQUENCE

When you know how many images you want to acquire and at what rate, a sequence acquisition is often appropriate. You can control the number of frames and which frames to acquire using IMAQ Sequence. In Figure 3.11, a number of images are created (in the for loop), and an array of those images is passed to IMAQ Sequence. Once IMAQ Sequence is called, it acquires the next number of frames as defined by the number of images passed to it. The time taken to acquire the frames is determined by the frame rate:

$$AcquisitionTime(s) = \frac{NumberOfFrames}{FrameRate(Hz)}$$

For example (Figure 3.12), a unibrain™ Fire-i400 1394 camera can acquire images at a frame rate of up to 30 frames per second. If you require 100 frames to be acquired, the acquisition would take:

$$AcquisitionTime = \frac{100}{30\,Hz}$$

$$\approx 3.3s$$

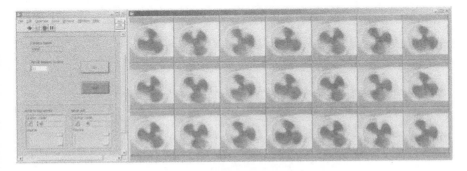

FIGURE 3.11 Simple sequence example – wiring diagram.

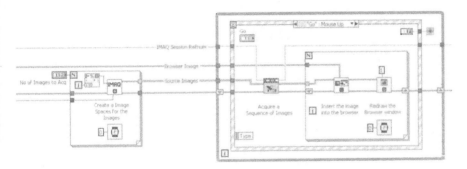

FIGURE 3.12 Simple sequence example.

IMAQ Sequence also has a *Skip Table* array input, which allows you to define a table of which frames to acquire and those to skip. The *Skip Table* array is defined as each element representing the number of frames to skip before acquiring each corresponding buffer. Therefore, the number of elements in the *Skip Table* should be the same as the number of elements in the *Images In* array. The number of skipped frames in the Skip Table should also be considered when calculating the acquisition time.

3.3 NI-IMAQ FOR IEEE 1394

Image acquisition is a little different when using an IEEE 1394 (FireWire) camera instead of a National Instruments frame grabber. To use a FireWire camera, you will either need to access third-party camera drivers, or use the National Instruments generic driver, *NI-IMAQ for IEEE 1394* (Figure 3.13). Installing *NI-IMAQ for IEEE 1394* enables access to FireWire cameras from MAX, and adds a subpalette to the *Motion and Vision* function palette (Figure 3.14).

The IEEE 1394 function palette contains VIs that replace the standard NI-IMAQ image acquisition VIs. All of the example VIs in this chapter have been replicated with IEEE 1394 capabilities on the accompanying CD-ROM (Figure 3.15).

FIGURE 3.13 NI-IMAQ for IEEE 1394.

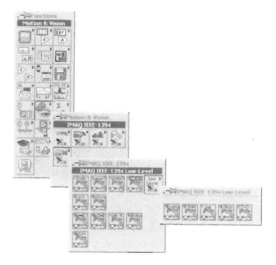

FIGURE 3.14 IMAQ IEEE-1394 Function Palette.

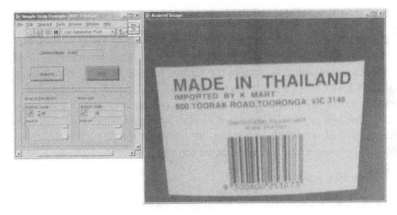

FIGURE 3.15 Simple snap example (IEEE 1394).

3.4 USER SOLUTION: WEBCAM IMAGE ACQUISITION

Peter Parente holds his B.S. in computer engineering and computer science from Rensselaer Polytechnic Institute (RPI). His interest in inexpensive image acquisition came about as part of his work with the Center for Initiatives in Pre-College Education (CIPCE) at RPI (see http://www.cipce.rpi.edu for more information). Peter is currently pursuing an additional degree in computer science with a focus on computer graphics at the University of North Carolina–Chapel Hill.

Version 1.4 of the Webcam VIs can be found on the CD-ROM accompanying this book. The most recent version can be found at http://www.mindofpete.org.

3.4.1 INTRODUCTION

Much of the image acquisition functionality of LabVIEW has remained out of the hands of the typical consumer because of the often prohibitively high prices of professional-grade image acquisition hardware. In many cases, the image quality provided by these expensive pieces of equipment goes far beyond the needs of LabVIEW programmers. Many LabVIEW applications can benefit from merely having images of a process displayed on the front panel, possibly with some simple image enhancement applied by the Vision Toolkit. In such cases, standard acquisition devices can seem overly complex, and definitely too pricey for their intended use (Figure 3.16).

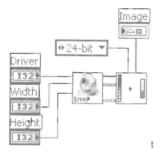

FIGURE 3.16 Single Capture with Display.

Several inexpensive solutions currently exist, but generally they are drivers for particular models and brands of cameras, and are often operating-system dependent. One attempt at adding such functionality appeared in RoboLab, a system developed by Lego, National Instruments, and Tufts University that allows users to program the Lego Robotics from within the LabVIEW environment. Version 2.5 of this application introduced support for the Lego Vision Command USB camera — a Logitech Quick-Cam housed in a special Lego casing. While RoboLab now allows the robotics hobbyist to capture and process image data using LabVIEW tools, it lacks support for many new versions of Windows, and often captures at a slow rate of about two frames per second at a resolution of 320 × 240 × 24. Additionally, the functionality is nested deep within the RoboLab software, proving less useful for those who wish to use the functionality directly within LabVIEW.

Another attempt has been made at filling the niche for LabVIEW support of inexpensive imaging hardware, and has so far proven to be quite successful. The open source LabVIEW Webcam library, released in July 2002, allows images to be captured on machines running nearly any version of Windows at speeds of 7 to 10 frames per second at $320 \times 240 \times 24$ resolution. The library relies on a simple dynamic link library (DLL) that communicates with the Video for Windows (VFW) programming interface in order to initiate, configure, begin and end a capture session with an imaging device. Labeling it a "Webcam library" is a slight misnomer, as VFW allows for communication with practically any imaging device with a fully supported Windows driver.

3.4.2 FUNCTIONALITY

The Webcam library consists of two sets of VIs built around the low-level function-ality of the DLL. The first set assists in the capturing of images using the DLL described previously. The second set provides a means of converting captured images to the IMAQ format for processing, to the LabVIEW picture format for display on the front panel, and to the bitmap format for file storage. When used together, the capture and conversion VIs provide a straightforward means of acquiring and work-ing with image data from consumer-grade devices. A number of example uses of the Webcam library are provided in the subsequent discussions.

3.4.2.1 Single and Continuous Capture with Display

Only two VIs are required to capture a single image from an imaging device and display it on the front panel. The Webcam *Snapshot VI* can be coupled with the Webcam *Flat to Picture VI* to accomplish this goal as shown in Figure 3.17.

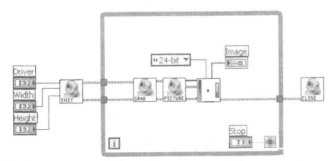

FIGURE 3.17 Continuous Capture with Display.

A slightly more complex configuration of VIs is required to capture images continuously. The Webcam *Snapshot VI* used in the single-shot acquisition shown in Figure 3.17 initializes a session, grabs a single image, and then immediately closes the session. A similar approach is taken when performing a continuous capture, except that the grab, conversion and display operations are repeated inside a loop, with the necessary initialization and discarding VIs outside, as shown in Figure 3.18.

3.4.2.2 Single Capture with Processing and Display

Any Webcam library image can easily be converted to the IMAQ format for image processing and analysis using the LabVIEW Vision Toolkit. The Webcam *Flat to IMAQ VI* is capable of creating a new buffer or using an existing IMAQ buffer to

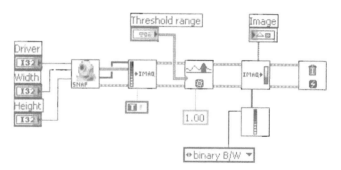

FIGURE 3.18 Single Capture with Processing and Display.

hold image data captured from an imaging device. The resulting IMAQ image can
be used with any of the Vision Toolkit VIs. Likewise, any processed images from
the Toolkit can be converted back into the picture format using the Webcam *IMAQ
to Picture VI* for display. The block diagram in Figure 3.19 demonstrates conversions
to and from the IMAQ format to perform a binary threshold operation on a single
captured image and display the result.

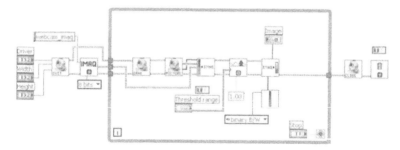

FIGURE 3.19 Continuous Capture with Processing and Display.

3.4.2.3 Continuous Capture with Processing and Display

Processing can also be done within a continuous capture session by including the
necessary VIs within the capture loop. The one notable difference is that the IMAQ
image is created outside of the loop and passed to the Webcam *Flat to IMAQ VI*.
The block diagram in Figure 3.20 demonstrates continuous image acquisition, processing and display loop.

3.4.2.4 Future Work

The Webcam library represents a solid step forward in providing robust, intuitive
and inexpensive image acquisition functionality in LabVIEW. However, the library
itself can still be improved by the addition of error reporting, and extended to run
on operating systems other than Windows. The project is under an open source
license, allowing all those who wish to improve or modify the library the freedom
to do so.

3.5 OTHER THIRD-PARTY IMAGE ACQUISITION SOFTWARE

In addition to Peter Parente's Webcam *Image Acquisition* suite, Irene He has developed a functional toolkit called *IVision*, which can acquire images from FireWire and USB cameras. IVision also contains VIs that control some of the camera settings, including focus, zoom, hue, saturation, brightness, frame rate and image size (providing your hardware supports these features). The software also allows reading and writing of several image file formats (BMP, DIB, JPEG, PNG, PBM, PGM, PPM, TIFF, and more), and simple image-manipulation functions. One of the IVision Toolkit's best features is the free availability of source code, so recompilation to previous LabVIEW versions is possible. More information about IVision (including a downloadable version) is at http://www.geocities.com/irene_he/IVision.html.

3.6 ACQUIRING A VGA SIGNAL

An interesting tidbit of information recently cropped up on the National Instruments Developer Zone: how to acquire a VGA video signal from a computer's video card output. When testing the way your software operates, it may be advantageous to inspect a screen shot of your application. If you attempt to acquire an image of your screen using a camera, many unwanted artifacts may be observed: glare from the screen, mismatched refresh rates causing distortion bars and limited process repeatability. It is possible to "tap" into the 15-pin VGA D-sub connector on the back of your video card and use an NI-1409 to acquire the video signal directly. The following video modes are supported using this technique:

- 640 × 480 70 Hz
- 640 × 480 60 Hz

More information (including the VGA connector wiring diagram, icd files and further instructions) can be found by visiting http://www.zone.ni.com and entering *"Using the IMAQ 1409 to Acquire a VGA Video Signal from a PC Video Card"* in the Search box.

3.7 TWAIN IMAGE ACQUISITION

TWAIN is an image capture API first released in 1992, and typically used as an interface between image processing software and a scanner or digital camera. The word TWAIN is from Kipling's *The Ballad of East and West* ("...and never the twain shall meet..."), reflecting the difficulty, at the time, of connecting scanners to PCs (a TWAIN acronym contest was launched, with the de facto winner being *Technology Without An Interesting Name*). Graftek and AllianceVision have released LabVIEW TWAIN drivers that allow direct access to TWAIN-compliant scanners and cameras, and import of images into LabVIEW and the Vision Toolkit. More information can be found at http://www.alliancevision.com/net/twain/fr_twain.htm.

3.8 USER SOLUTION: COMBINING HIGH-SPEED IMAGING AND SENSORS FOR FAST EVENT MEASUREMENT

Bryan Webb
Marketing Manager
Microsys Technologies Inc.
Tel 905 678-3299 x3600
Email: bryanwebb@micro-sys.com

3.8.1 THE CHALLENGE

A series of customers needed systems to control automotive safety for short duration airbag and seatbelt pretensioner tests. The systems required high-speed cameras and data acquisition, and needed to generate various control signals. Customers also asked for the ability to view and analyze the collected test data. The challenge was to design and construct a highly integrated and automated product based on standard off-the-shelf hardware that could be easily configured in the field for the different applications and tests.

3.8.2 THE SOLUTION

National Instruments LabVIEW software was used to control NI-DAQ, NI-DMM and Field Point hardware in addition to the high-speed imagers and other ancillary equipment and safety interlocks. A companion program (also written in LabVIEW) provides sophisticated viewing and analysis capability for both image and sensor data.

The automotive market requires the ability to design and build thousands of parts that go together to construct a car. Consumers have embraced safety as a saleable concept, and manufacturers want to provide products that fill that need. In addition, the U.S. government mandates specific tests and pass-performance levels that must be observed. Original equipment manufacturers have outsourced some design work to Tier 1 and Tier 2 companies such as Autoliv, TRW, JCI, Magna/Intier, Delphi and Visteon, who supply parts and perhaps complete systems or interiors to companies higher up the production chain. This process has reduced in-house test staff in many companies, often leading to the purchase of commercial test systems rather than building complex one-off systems in-house, where learning curves can be long, steep and therefore costly.

The stages of a product's life include *engineering* (design validation, total performance testing) and *production* (product verification, lot acceptance testing, or conformity to production). Overall, testing helps validate appropriate designs, verify appropriate production, and thus limit waste and liability. The question is, "what to test?" For example, one could simply perform an electrical or physical inspection on a crude airbag deployment (without any triggered video or data monitoring) just to see if the bag deploys. However, there is still a question of determining whether it deploys correctly, as per the product specification.

FIGURE 3.20 The Operations Console.

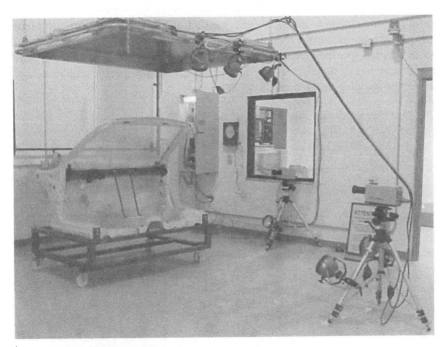

FIGURE 3.21 A Typical SureFire Airbag Test System.

To meet these challenges, Microsys introduced SureFire™ (Figure 3.20), a fifth-generation automated test system based on industry-proven National Instruments LabVIEW software, data acquisition (DAQ), FieldPoint and digital multimeter (DMM) hardware. SureFire is capable of performing a comprehensive list of airbag deployment tests (with multiple squibs), seatbelt pretensioner tests, head impact tests (drop tower), and handling data in dynamic crash (high-G sled or barrier) applications (Figure 3.21). PowerPlay™ provides powerful analysis tools with synchronized high-speed imager and sensor data displays coupled with image enhancement functions and sensor filters and mathematical functions. To complete the test, PowerPlay supports the creation of multiple image movies in AVI format for distribution and Excel™-based test reports.

3.8.3 SYSTEM CONSIDERATIONS

To ensure that the system can handle different test needs, a modular software approach is used for hardware and other capabilities, which also minimizes the integration time and effort for future developments. SureFire calls hardware driver modules as they are required to perform different functions. A similar approach is used for other features that may be needed by one application and not another (e.g., automatic AVI creation at the end of a test). New hardware modules can be developed and added to SureFire to meet changing test demands. The system can configure and control almost any high-speed imager by loading the appropriate driver. For example, if the customer needs to perform squib resistance measurements (Figure 3.22), the appropriate driver module (and hardware) is added. The test system-type and application-specific demands dictate the driver modules that accompany Sure-Fire. This allows us to tailor each system to the customer's specific requirements at the time of shipping and for later system upgrades.

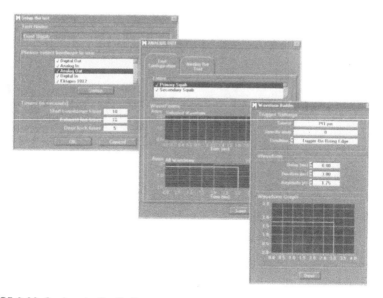

FIGURE 3.22 Setting the Squib Outputs.

Each SureFire Test System installation has its own unique combination of hardware — variable quantities and types of safety interlocks, climactic chamber controls, imager systems, and sensor channels — which all require system interfacing and control. It is important in any test scenario to produce repeatable and reliable results that lead to customers being able to make informed business decisions. One of our key design objectives was to minimize the time between order and delivery of the system, while keeping system costs in line with customer budgets by using industry-standard components. The powerful graphics programming available in LabVIEW (for DAQ, DMM, and FieldPoint components) was critical to the success of the system. These components provided the required flexibility and minimized expensive engineering time. Since 1992, the system has been evolving based on customer demands and feedback.

3.8.4 SYSTEM PERFORMANCE

The system was designed to dynamically configure its hardware and to control and monitor all of the events in a specified test sequence, which the customer can design and modify without the assistance of Microsys engineers. For the most part, each system is based on a standard platform, which is tailored to meet the specific customer requirements. During installation, the software is initially configured with customer specifics, including hardware modules, data entry field names, and sensor data and calibration information.

One of the unique features of SureFire is its integrated high-speed video image and DAQ data control capability. The ability to configure and control various high-speed video imaging and DAQ equipment from a single computer simplifies test preparation and execution, reducing user error and improving test efficiency. Fewer tests are performed erroneously, leading to less scrap and lower testing costs.

SureFire is database driven (either Microsoft Access or Oracle), therefore all test configurations, hardware calibration information and test-result pointers are stored for each test. This provides management with a tool for traceability.

For analysis, PowerPlay provides integrated video and DAQ displays to give engineers the ability to easily correlate high-speed video images with sensor data in a fashion similar to National Instruments DIAdem. Users can also compare and evaluate multiple test results by displaying both video and sensor data from different tests. PowerPlay will play back all open data windows in a time-synchronized fashion so that engineers can compare and confirm visual clues with sensor data. Image enhancements include linear and angular measurements, color correction, sharpen and blur filters and edge detection and enhancement.

3.8.5 APPLICATIONS

Other applications for SureFire include aerospace and defense, testing products such as ejection seats, landing shocks, missile launch (ground-based tests) and ordnance. SureFire is flexible enough to meet many applications where there is a need to integrate and control high-speed imaging and sensors with test control. PowerPlay can be used as a stand-alone analysis tool to combine computer-generated images with real test images to tweak a simulation model. PowerPlay can import and export to data files such as DIAdem, CSV, ISO13499, TIFF, JPEG, BMP, AVI and GIF.

3.8.6 SUMMARY

SureFire performs common automotive safety tests through its modular hardware and software design. Each test system is tailored to meet customer requirements for safety interlock control and monitoring, auxiliary equipment control, test trigger, high-speed video capture, and sensor data acquisition. Integrated high-speed video and DAQ hardware coupled with image enhancement and math functions provide users with advanced test analysis capabilities.

4 Displaying Images

Although not always necessary, displaying acquired and processed images to the user is often desirable. If your application calls for the user to be able to review image data, it is wise to consider performance and user interface designs very seriously. Your first question should be, "does my application *really* call for image data display?" Although we all like to "show off" our programming expertise, and there is no better way than to show the users (read: managers) the before-and-after data, displaying images can have much more detrimental effects to system performance than traditional DAQ systems.

First, image data can often contain much more information than simple DAQ data — a voltage trace with respect to time may be only an array of a few thousand doubles, whereas a similar application acquiring images over time could easily fill up your RAM (e.g., a 1024×768 32-bit image is approximately 3 Mb). Similarly, image data generally contains more "background" data than tradition methods, simply due to its two-dimensional nature.

Second, displaying any data on your front panel will take time and memory. Using a multithreaded version of LabVIEW, whenever you wire data to an indicator (numeric, string, Boolean, image, etc.) the *execution* thread is stopped, and the system switches to the *user interface* (UI) thread. The UI thread's copy of the indicator's data is then updated (this is generally called the *operate* data), and if the panel is open, it is redrawn. Then the data is sent back down to the *execution thread*, but in order to do this, it is first sent to a protected area of memory called the *transfer buffer* with a *mutual exclusion object* (MUTEX). After all of this is done, the systems switches back to the execution thread, and when it next reads data from terminals, it finds any new copies in the transfer buffer and uses these as the new values. As you can imagine, copying image data between these layers can be quite time and RAM consuming, especially if the dataset is large or inappropriately formatted.

You need to decide if you are willing to trade off system execution speed for image display — if you do not really need to display an image on your front panel, I suggest that you do not (perhaps create a modal sub-VI that the user can view only when it is required). On the other hand, you may not be able to get away from telling the user what is going on, but you will need to think carefully about the required update rate and how you are going to transfer the data.

4.1 SIMPLE DISPLAY TECHNIQUES

If you are reading this book, then you probably have already installed the NI Vision toolkit, and perhaps even played around with some of the example VIs that come with it. Figure 4.1 is a simple example that you might recognize (it has been slightly modified from its original functionality). As you can see from the comment text in

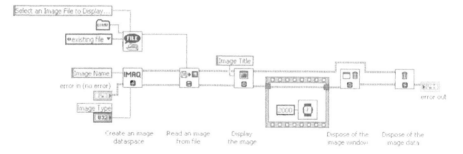

FIGURE 4.1 Simple image display — wiring diagram.

FIGURE 4.2 Simple image display.

the diagram, this example allows the user to select an image to load from disk, and displays it in its own floating window (Figure 4.2).

This technique is very simple to implement, and displaying multiple floating windows is only a little more complex. The IMAQ WindDraw VI has a window number input, allowing you to specify which floating window you want this operation to be performed on. By default, this is set to zero, but you can specify up to 16 simultaneous windows. Figure 4.3 demonstrates five simultaneous windows. This is achieved by looping around both IMAQ WindDraw and IMAQ WindClose (Figure 4.4).

Figure 4.1 and Figure 4.4 both use IMAQ WindClose and IMAQ Dispose to clean up the floating windows and release the image data from memory. Technically, you can leave the floating windows open, even after your VI has finished executing. However, this is bad form. Doing so may lead to poor memory management, with sections of the RAM continuing to be locked and therefore unusable after your application has quit, forcing the user to manually close them. When you dispose an image that is still displayed, if you minimize or drag another window over it, the system will lose the information of the image, and will not be able to refresh the screen correctly. Rather than dispose of each window separately, you can set the

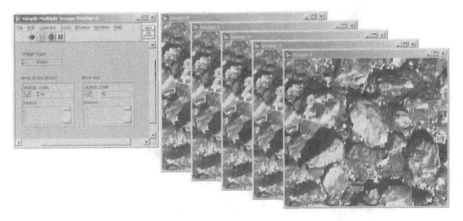

FIGURE 4.3 Simple multiple image display.

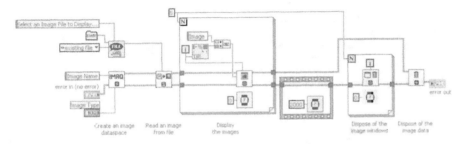

FIGURE 4.4 Simple multiple image display — wiring diagram.

IMAQ WindClose terminal *"Close All Windows?"* to true, and all windows currently open (including any embedded windows, as discussed later) will close.

4.2   DISPLAYING IMAGES WITHIN YOUR FRONT PANEL

Although displaying your image in floating windows can be useful, a common question asked by beginner Vision Toolkit programmers is, "how do I embed a dynamic image in my front panel?" There are several methods of doing so, depending on what you want to display and the speed restrictions you are willing to accept.

4.2.1   THE INTENSITY GRAPH

The simplest method of displaying an image on a VI's front panel is to convert it to an *intensity graph* (Figure 4.5). This method uses IMAQ ImageToArray (Figure 4.6) to convert the image data to an array that the intensity graph can plot. Do not forget to either autoscale the z axis (intensity), or set its upper limit to an appropriate value (the default is 100, whereas you will probably need to set it to 255, as shown in Figure 4.5), otherwise your intensity data will be clipped.

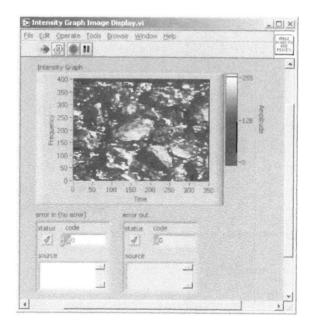

FIGURE 4.5 Intensity graph image display.

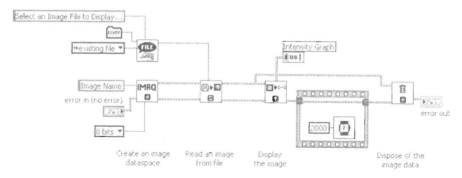

FIGURE 4.6 Intensity graph image display — wiring diagram.

4.2.2 THE PICTURE INDICATOR

Although using the intensity graph can be quite fast, it is limited to displaying 8-bit images — if you would like to display other image types, you can convert the image data array into a picture, and subsequently display it in a picture indicator (Figure 4.7)

You can use Draw 1-bit Pixmap, Draw 4-bit Pixmap, Draw 8-bit Pixmap, Draw TureColor Pixmap and Draw Flattened Pixmap to convert the appropriate data array to a picture (Figure 4.8). As you can probably imagine, this process of first converting the image data to an array and then to a picture can be very slow, depending on the image type, especially the step from an array to a picture. For this

FIGURE 4.7 Picture control image display.

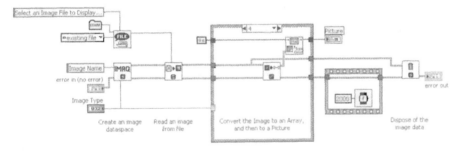

FIGURE 4.8 Picture control image display — wiring diagram.

reason, displaying images in picture controls is not generally appropriate for frequently updating data sets (e.g., displaying video sequences).

You need not use the IMAQ functions to load and display JPEG, PNG or BMP images; the LabVIEW Base Development System comes with VIs to accomplish this without installing the Vision Toolkit (Figure 4.9).

Use the appropriate VI from the Graphics&Sound>GraphicsFormats function palette (Read JPG File, Read PNG File or Read BMP File) to return the file as flattened image data, and then convert it to a picture using Draw Flattened Pixmap (Figure 4.10). Unfortunately, if the picture indicator is not the correct size, the image will either be clipped or will not fill up the entire front panel object. This problem can be avoided by resizing the indicator to fit the loaded image (Figure 4.12). Resizing

FIGURE 4.9 Picture control image display (without IMAQ).

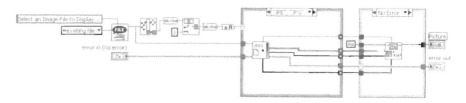

FIGURE 4.10 Picture control image display (without IMAQ) — wiring diagram.

front panel objects can have undesired side effects. The picture indicator may become too large for the front panel, and even overlap other front panel objects, as shown above (Figure 4.11).

You can also use this non-IMAQ Vision Toolkit technique to display images dynamically from Web addresses. A randomly chosen CRC Press Web page shows an image of a book cover, which can be displayed in a Picture indicator on a front panel (Figure 4.13). By using `DataSocket Read`, the raw ASCII text of the image file is read, and subsequently saved to a temporary text file on the local hard drive. Then using the techniques described earlier, the file is opened, and the unflattened data converted to a picture (Figure 4.14).

FIGURE 4.11 Picture control image display (without IMAQ) — autosize.

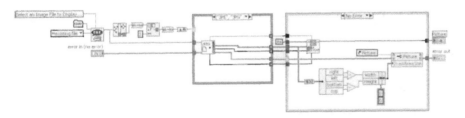

FIGURE 4.12 Picture control image display (without IMAQ) — autosize — wiring diagram.

4.2.3 EMBEDDING AN IMAQ DISPLAY WINDOW

The IMAQ display window is by far the fastest way to display images, but having the data in a floating window may not always be appropriate. Embedding a Vision floating window in your front panel is also possible, but unfortunately the method of doing so is not particularly intuitive (Figure 4.15).

This technique uses the FindWindowA and SetParent prototypes of the user32.dll Windows® library to "attach" the floating daughter to the nominated window. Before calling this VI, the parent VI must be loaded into memory. You should then spawn the floating window, set its properties (including moving it to

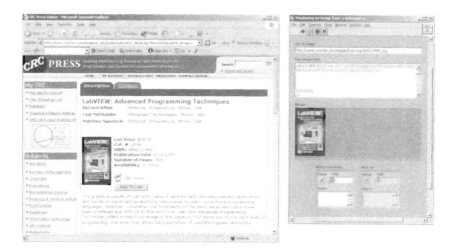

FIGURE 4.13 Displaying an image from a web page.

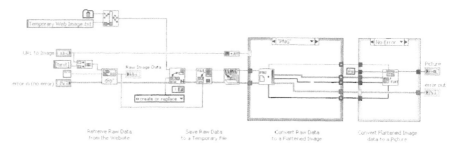

FIGURE 4.14 Displaying an image from a web page — wiring diagram.

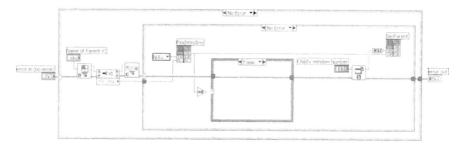

FIGURE 4.15 Embed an IMAQ Vision window in a parent — wiring diagram.

the correct position relative to the parent), and then run this VI to execute the embedding (Figure 4.16). You are also able to embed the child first, and then move it to its correct position, decreasing the time that the floating child is displayed. This second method is usually preferred, as subsequent operations on the embedded child are relative to the parent; for example, if you embed the child, and then execute an

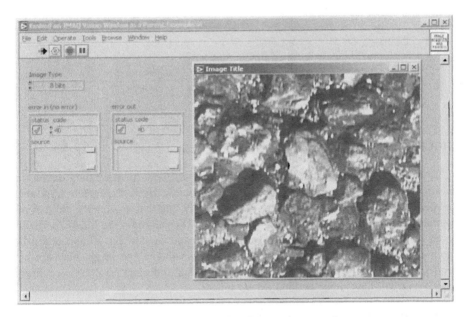

FIGURE 4.16 Embed an IMAQ Vision window in a parent example.

IMAQ WindMove, the coordinates that you enter are now relative to the upper left corner of the parent VI (including its title bar, toolbar and menu, if any).

National Instrument's Greg McKaskle recently posted a message to the Info-LabVIEW mailing list to help clear up questions related to some confusing Windows operating systems terminology (for the complete transcript of the post, see http://messages.info-labview.org/2002/10/14/10.html):

> Each window can have a parent and an owner.... Each window in LabVIEW is owned by the LabVIEW taskbar button's window — which is hidden by the way. Typical top-level LabVIEW panel windows do not have a parent, otherwise they would be contained (clipped and moved relative to the parent). Again, they do have an owner, but this doesn't imply anything about their position, or their clipping. The owner is used to send messages to, primarily during key processing.

Although it may look as though the child window is floating on top of the parent VI in the previous example, it is actually embedded; dragging, minimizing and resizing the parent affects the embedded child as though it is part of the parent's front panel (you can easily remove the child's title bar by calling IMAQ WindSetup and setting *"Window has Title Bar?"* to false). Although the embedded child may seem to be part of the parent, it is not. LabVIEW does not know (or care) that the child has been embedded, and does not treat it any differently. Therefore, it is still important to remember to dispose of both the embedded child window and image data properly (Figure 4.17).

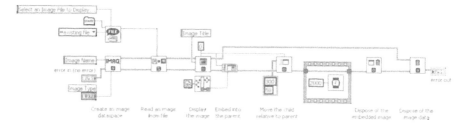

FIGURE 4.17 Embed an IMAQ Vision Window in a parent example — wiring diagram.

Although the child window is embedded in the parent, programming data to it is almost no different to when it was a floating window. However, keep in mind the relative spatial coordinates as mentioned previously, and that moving the child even partially outside of the parent's bounds (Figure 4.18) will mask the image (perhaps even completely, where it may seem that your image window has disappeared). If you scroll a front panel while it contains a child window, the image may not be correctly redrawn. Therefore, it is often a good idea to disable scrolling on a front panel that will have an embedded IMAQ image.

4.3 THE IMAGE BROWSER

The image browser is a very useful function if you need to display a selection of thumbnails to a user, perhaps with them selecting one (or more) for further processing. The image browser itself is displayed as a floating IMAQ image window (Figure 4.19).

Loading images into the image browser is quite simple. First, the browser is configured, including its background color, size, style, and the number of images you want displayed per line. The image browser is then displayed, like any other image, and populated with images. Figure 4.20 shows the image browser is updated after each image is inserted into it, but you could put the IMAQ WindDraw outside of the for loop, causing the program to wait until all of the images have been inserted, thus displaying them all simultaneously.

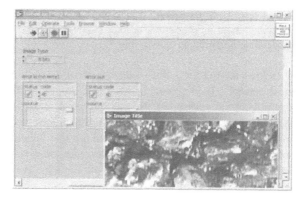

FIGURE 4.18 Embedded window poor move example.

FIGURE 4.19 Simple image browser example.

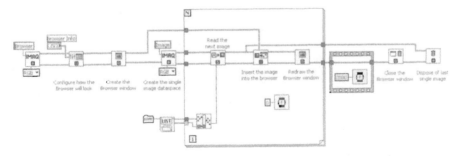

FIGURE 4.20 Simple image browser example — wiring diagram.

As the image browser is a floating **IMAQ** image window, you can embed it as a child into a parent **VI**, as discussed previously (Figure 4.21). Adding this code embeds the image browser into the parent VI's front panel (Figure 4.22).

One useful example of image browsers is to allow the user to select a thumbnail, and then display (and perhaps process) the full image on the front panel of a VI. This is achieved by embedding both the image browser and preview image into the parent VI, and using the selected browser thumbnail to dictate which image is loaded into the preview image (Figure 4.23). As not to continually load images into the preview image, even when the user has not selected a different image, IMAQ WindLastEvent is used to monitor if the user has clicked within the image browser (Figure 4.24)

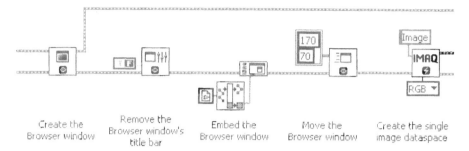

Create the | Remove the | Embed the | Move the | Create the single
Browser window | Browser window's | Browser window | Browser window | image dataspace
 | title bar | | |

FIGURE 4.21 Simple embedded image browser example — wiring diagram.

FIGURE 4.22 Simple embedded image browser example.

FIGURE 4.23 Image browser-selector example.

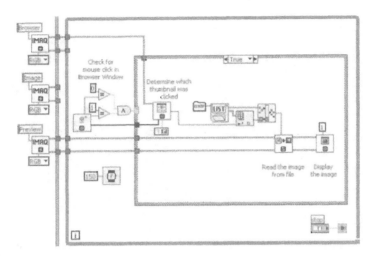

FIGURE 4.24 Image browser-selector example — wiring diagram.

4.4 OVERLAY TOOLS

A picture may be worth a thousand words, but those words are often open to the viewer's interpretation. You can give the user an enormous amount of indicative information by overlaying text and other types of pictorial data over an image.

Overlaying is exactly that — placing information on top of the image. It does not replace any of the image data (all the underlying pixels remain intact), and overlays can be moved, changed or even removed completely with no effect on the underlying image.

4.4.1 OVERLAYING TEXT

You can overlay text to describe an image to the user and provide technical, customer and company information. The fonts available to IMAQ Overlay Text are limited only to those installed on the local machine (the example in Figure 4.25 uses *Arial*), although some of the font options (strikeout, italic, underline, outline, shadow and bold) may not be available for the font you have selected. You should remember to include any nonstandard fonts that you have used when distributing your VIs, as they may not be installed on the target machine. If the VI cannot find a user-specified font, it will default to the last font successfully selected — often with unimpressive results.

Simply supply the text, its color (you can also specify its background color but the default is *transparent*), position coordinates and font attributes to the VI and your text will be overlaid (Figure 4.26).

4.4.2 OVERLAYING SHAPES

You can demonstrate several points or regions of interest including spatial measurements, using shape overlays. The default shapes include circles, rectangles and arcs (Figure 4.27). Each of the shapes have definable colors, drawing bounds and fill

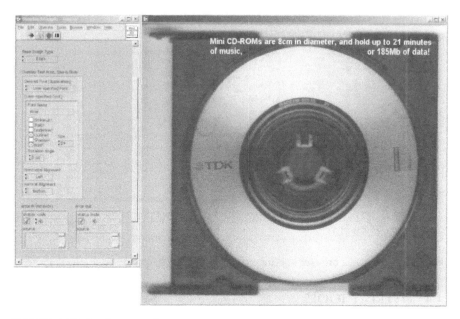

FIGURE 4.25 Overlay example — text.

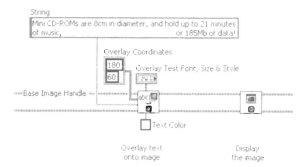

FIGURE 4.26 Overlay example: text — wiring diagram.

options (either *fill* or *frame*), and the arc shape overlay also requires start and end angles (Figure 4.28).

4.4.3 OVERLAYING BITMAPS

Bitmap overlaying can be useful to display a company logo, a data graph image, a custom coordinate scale, or just about any other bitmap over the base image. The example in Figure 4.29 shows the overlaying of a company logo. The image input of the IMAQ Overlay Bitmap VI is unfortunately in the form of a U32 array: a standard image input could have made life a little easier when using this function. This means that all images must first be converted to an array before they can be overlaid. Once

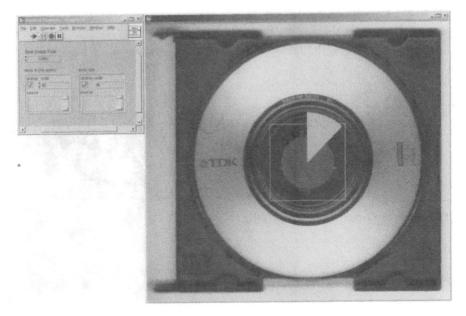

FIGURE 4.27 Overlay example — shapes.

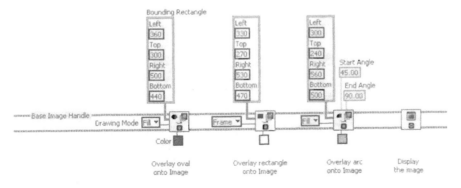

FIGURE 4.28 Overlay example: shapes — wiring diagram.

you have converted your image, you can also define the positional coordinates of the overlay, relative to the top left corner of the base image (Figure 4.30).

4.4.4 COMBINING OVERLAYS

As you can see, you can overlay multiple objects on a Vision window, as well as different types of overlays. Figure 4.31 is an example of overlaying all of the types discussed previously. Combining overlays allows a highly configurable, dynamic method for displaying information on images, perhaps for user information, or inclusion in a printed report.

FIGURE 4.29 Overlay example — bitmap.

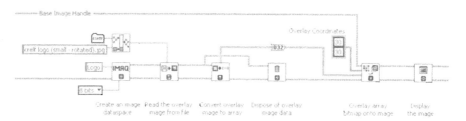

FIGURE 4.30 Overlay example: bitmap — wiring diagram.

FIGURE 4.31 Overlay example — all.

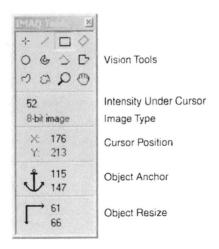

FIGURE 4.32 Vision tools palette.

4.5 THE VISION WINDOW TOOLS PALETTE

If you need to do more than just display an image, and you are looking for some user interaction, one often overlooked tool for the job is the *Window Tools Palette* (Figure 4.32). This floating palette allows the user to select one of many tools to measure data, draw objects, zoom, pan and easily define regions of interest (ROIs), while providing feedback such as image type and current relative mouse coordinates. You display and hide the tool palette using the IMAQ WindToolsShow VI, which works in a similar manner to IMAQ WindShow (Figure 4.33). Even before you have displayed your tool palette, you can programmatically alter its functionality (Figure 4.34). As you see, the palette is very customizable using IMAQ WindToolsSetup and IMAQ WindToolsMove, allowing you to define which tools to display, the position of the palette, whether to show mouse coordinates, and how many tools are displayed per line (Figure 4.35).

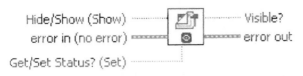

FIGURE 4.33 IMAQ WindToolsShow.

4.5.1 AVAILABLE TOOLS

There are 13 tools available on the Vision Tools Palette (Figure 4.36):

Selection Tool (Tool 0): The selection tool allows the user to select a previously defined ROI or shape on the image.

Point (Tool 1): The point tool allows the user to select one pixel, and can be useful to nominate previously defined shapes, or return information particular to that pixel (intensity or RGB values, for example).

FIGURE 4.34 Vision window tools palette — select tools.

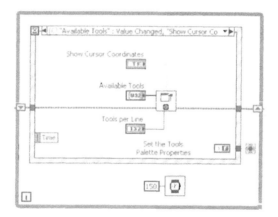

FIGURE 4.35 Vision window tools palette: select tools — wiring diagram.

FIGURE 4.36 Vision window tools palette tools.

Line, Rectangle and Oval (Tools 2, 3 and 4): These tools are similar to the *point tool*, except that they offer one- (line) and two-dimensional (*rectangle* and *oval*) area selection. Figure 4.37 shows an inverse mask applied to several area selections, including a *rectangle* and *oval*.

FIGURE 4.37 Vision window tools palette — multiple tools demonstration.

Polygon (Tool 5): The example above includes a demonstration of the *polygon* tool — the n-sided polygon begins with a left mouse click, and each subsequent side is defined by the next mouse click. The polygon is then closed to the first point with a double click.

Free (Tool 6): As its name suggests, this tool allows the user to draw a freehand flat-sided region in the image. The first point is defined by left clicking, as are subsequent vertices. Double clicking will close the ROI.

Annulus (Tool 7): An annulus is the area between two concentric sliced circles. This tool is useful when analyzing parts like spanners, and terminal clamps. The two circles are initially created with a left click drag (first click defines the center of the circles, and the second defines the radius of the outer circle), and then the slice and radius of the inner circle is defined by dragging the appropriate ROI handles.

Zoom and Pan (Tools 8 and 9): These tools allow the user to manipulate the way an image is displayed, by either magnifying a portion of the image, to moving within the image. Zoom and pan are very useful when the user needs to observe small areas of a large image, zooming in and then moving the viewable area to adjacent areas.

Broken and Free-Hand Lines (Tools 10 and 11): As you can see from Figure 4.38, broken and free lines allow the user a free way of defining open and closed ROIs. The broken line creates a line as the user draws (click and drag to create the region), which is useful to trace features in the image, whereas the free line joins the last point defined with the first, creating a closed region.

Rotated Rectangle (Tool 12): The rotated rectangle tool is very useful when a rectangular ROI is required, but the part's acquired image is not exactly aligned with the coordinate system of the image acquisition system. The user can first draw the rectangle, and then rotate it by moving one of the ROI's rotation handles.

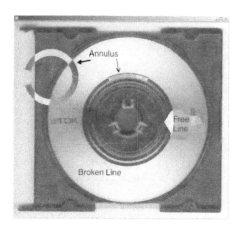

FIGURE 4.38 Vision tools palette — more tools.

4.5.2 VISION TOOLS PALETTE HINTS

While using the region tools, you can alter their functionality by using a tool by pressing certain keyboard keys, much like the selection tools when using Windows Explorer or Microsoft® Word. If you hold down *Shift* while drawing a ROI, it constrains the region's Δx and Δy dimensions to be equal, forcing free-hand rectangles to form squares, and ellipses to circles. Interestingly, if you hold down the *Shift* key when drawing a line segment, it will be forced to be either horizontal or vertical, and also constrains the rotation of rectangles and annuluses to multiples of 45° from horizontal.

As is common with Windows® applications, holding down the *Control* key while drawing elements on an image allows you to create multiple instances. To erase a previously created element, either select it and press the *Delete* key, or simply create a new element without holding down the *Control* key.

5 Image Processing

Once you have acquired images, they are often not in an appropriate format for you to analyze and must be processed first. The Vision Toolkit contains many tools to help you clean up and alter images, from defining a region of interest to filtering and spatial manipulation.

5.1 THE ROI (REGION OF INTEREST)

A ROI is a specification structure that allows for the definition of arbitrarily shaped regions within a given image, often called subimages (although a ROI can encompass the entire image if so defined). A ROI contains no image data — it is not an image itself, but a placeholder that remembers a defined location within an image.

Why would you need a ROI within an image? Why not just acquire the region of interest as the complete image? There are many answers to this question: first, ROIs can be irregular shapes — not necessarily the rectangle or square that your camera supplies. ROIs can be rectangles or squares (a square is often called an *area of interest*, or AOI), circles, annuluses, polygons and freehand shapes. Secondly, you may want to perform different image processing routines on different areas of an image — there is no sense in wasting valuable time on processing the whole raw image, when you are only interested in a few small parts of it. Another reason for ROIs is that you can dynamically define and change them — users can draw a ROI over a portion of an image at run time, selecting their particular region of interest.

The license plate in Figure 5.1 has a defined ROI around the alphanumeric characters, to make it much simpler and faster for an optical character recognition (OCR) routine to decode. Also, the routine would try to recognize characters where none are present, perhaps returning erroneous results.

FIGURE 5.1 Car license plate with a defined ROI

By traditional definition, every point of an image is either inside or outside of a given ROI. The Vision Toolkit takes this definition one step further by adding two new ROI concepts: the line and the point. Strictly speaking, these ROI types are not actually lines and points, but a single pixel width rectangle and square respectively. Points are useful to return single value data about a region of your image (point intensity, color data, etc) and the line can return one-dimensional information.

5.1.1 Simple ROI Use

5.1.1.1 Line

A simple ROI example is that of an intensity contour measurement, as shown in Figure 5.2 and Figure 5.3. The IMAQ WindLastEvent VI is used to detect the event that last occurred in the nominated IMAQ window and looks for a draw event (suggesting that the user has drawn a ROI of some type). IMAQ WindGetROI then returns the definition of the created ROI and IMAQ ROI Profile returns the intensity contours along the length of the ROI (if a two-dimensional ROI is defined, the intensity contours follow the outer shape of the ROI, starting at the first defined point).

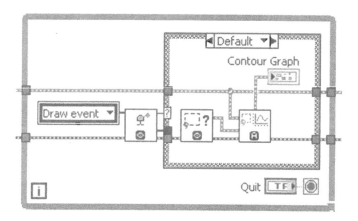

FIGURE 5.2 Simple ROI intensity contour — wiring diagram.

As you can see, the initial ROI point is that on the far left and as the line is followed, the intensity levels are returned (suggesting that the background intensity of the license plate is approximately 240 units, whereas the alphanumeric characters are approximately 50 units).

5.1.1.2 Square and Rectangle

A square ROI is the simplest two-dimensional ROI — it allows the user to select a region with four equal sides, although the rectangle ROI is probably the most popular, allowing the user more flexibility in defining the required region.

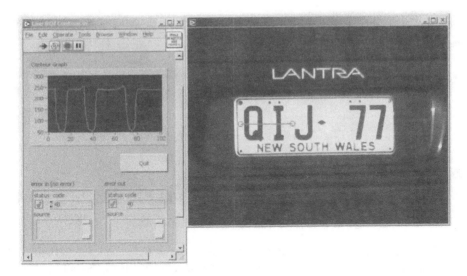

FIGURE 5.3 A line ROI, with its calculated intensity contour.

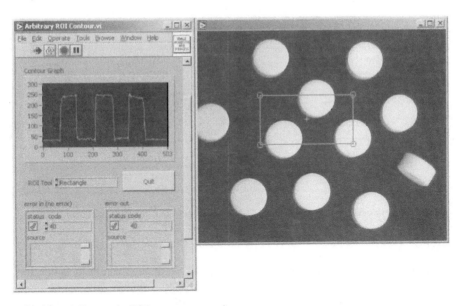

FIGURE 5.4 Rectangle ROI contour example

Figure 5.4 shows intensity contour information for a rectangular ROI — the contours are plotted from the first defined corner of the ROI (the top left corner in this case), along the ROI in a clockwise direction. As they are two-dimensional ROIs, they have several other applications, including image portion selection (See Figure 5.5).

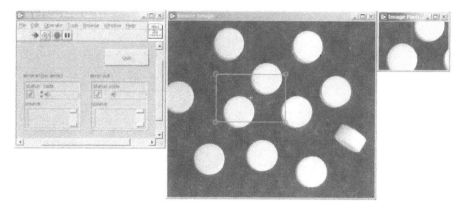

FIGURE 5.5 Two-dimensional ROI image portion selection.

Image portion selection is achieved by first converting a detected ROI to a rectangle and then extracting the portion of the source image that is bound by that rectangle using IMAQ Extract, as shown in Figure 5.6.

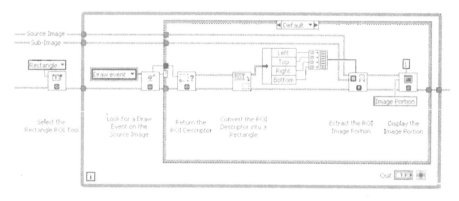

FIGURE 5.6 Two-Dimensional ROI image portion selection — wiring diagram.

5.1.1.3 Oval

An oval ROI is defined by first holding down the left mouse button to define the center and then dragging to define the radii of the ROI:

As you can see from Figure 5.7, the position of the mouse cursor during the drag operation may not be particularly intuitive, but it is a corner of the oval's bounding rectangle.

5.1.2 COMPLEX ROIs

Although simple ROIs can be very useful in particular situations, several other types are available, including the rotated rectangle and annulus arc.

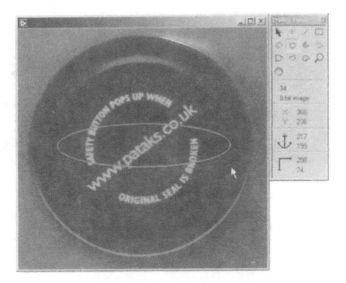

FIGURE 5.7 Oval ROI definition.

5.1.2.1 Rotated Rectangle

As its name would suggest, rotated rectangles are similar to their simple counterparts, but with the added functionality of allowing the user to rotate them. When a rotated square or rectangle is defined, several extra construction lines are evident (Figure 5.8).

The masked rotated rectangle ROI in Figure 5.8 was achieved with the code shown in Figure 5.9.

5.1.2.2 Annulus Arc

An annulus is defined as the figure bounded by and containing the area between two concentric circles and an annulus arc is, as you might expect, an arc component of that annulus (Figure 5.10).

An annulus arc is very useful when unwrapping text, or any otherwise wrapped image data, for analysis. Defining an annulus arc takes place in four steps, as shown in Figure 5.11(a) defining the outer radius, Figure 5.11(b) the starting edge, Figure 5.11(c) the ending edge and Figure 5.11(d) the inner radius.

Using the Annulus Arc selection tool alone may seem of little or no use, if it is combined with an unwrapping routine (IMAQ Unwrap, for example) you can unwrap a semicircular portion of an image, distorting it to a rectangular shape, as shown in Figure 5.12. This example waits for a Draw Event on either of the IMAQ windows and once one is detected, that window's ROI is returned, converted from a raw ROI to an annulus arc and then IMAQ Unwrap is called to unwrap the annulus arc into a rectangle.

The unwrapped image (the floating window containing the text in Figure 5.13) can then be fed through optical character recognition (OCR) software, including the

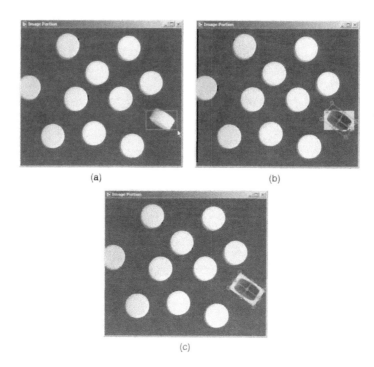

FIGURE 5.8 (a) Drag out a standard rectangular ROI. (b) Rotate the ROI with the handles. (c) Rotated rectangular ROI.

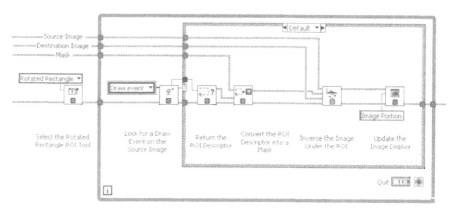

FIGURE 5.9 Rotated rectangle ROI image portion selection — wiring diagram.

National Instruments OCR Toolkit add-on for the Vision Toolkit (see Chapter 8 for more information about OCR).

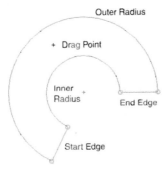

FIGURE 5.10 Annulus arc components.

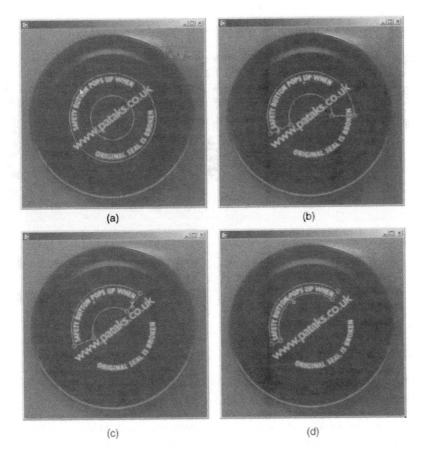

FIGURE 5.11 (a) Defining the outer radius. (b) Defining the starting edge. (c) Defining the ending edge. (d) Defining the inner radius.

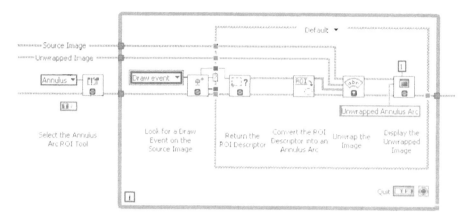

FIGURE 5.12 Unwrap annulus arc ROI — wiring diagram.

FIGURE 5.13 Unwrap annulus arc ROI.

5.1.2.3 ROI Tracing Example

A user recently posted a question on the National Instruments Developer's Zone discussion forum:

> I am currently working on a project using LabVIEW 6.0 and IMAQ Vision [and] I need to get the ... coordinates of the cursor in an IMAQ window...

The project called for the user to drag the mouse pointer over an image and then track a laser beam along the same coordinates. Although previous examples in this book have used IMAQ WindLastEvent to determine only the type of event that has occurred, it can also provide us with some limited information regarding the event's attributes. The VI has three outputs of interest: *event, coordinates* and *other parameters,* which are populated as shown in Table 5.1.

TABLE 5.1

Event	ROI Tool	Coordinates [Array]	Other Parameters [Array]
None	Any	NULL	NULL
Click	Cursor	[0,1] xy position of the click	[0,1,2] selected pixel intensity
Double click	Any	[0,1] xy position of the click	[0,1,2] selected pixel intensity
	Zoom	[0,1] xy position of the click	[0] new zoom factor
		[2,3] image xy center position	
Draw	Line	[0,1] xy start point	[0] bounding rectangle width
		[2,3] xy end point	[1] bounding rectangle height
			[2] vertical segment angle
	Rectangle	[0,1] xy start point	[0] width
		[2,3] xy end point	[1] height
	Oval	[0,1] bounding rectangle start point	[0] bounding rectangle width
		[2,3] bounding rectangle end point	[1] bounding rectangle height
	Polygon	[0,1] bounding rectangle start point	[0] bounding rectangle width
		[2,3] bounding rectangle end point	[1] bounding rectangle height
		[4,5], [6,7], [8,9] ... vertices	
	Freehand	[0,1] bounding rectangle start point	[0] bounding rectangle width
		[2,3] bounding rectangle end point	[1] bounding rectangle height
		[4,5],[6,7],[8,9]... vertices	
Move	Any	[0,1] new position of the window	NULL
Size	Any	[0] new width of the window	NULL
		[1] new height of the window	
Scroll	Any	[0,1] new center position of image	NULL

In this instance, the user was able to use the *draw vertices* data returned in the *coordinates* array, which was subsequently built into an array of clusters (each containing the *x,y* pair of the pixels that were traced over). This allowed the user to trace the path of the mouse (Figure 5.15).

As you can see in Figure 5.14, the *coordinates* array returns the start and end point of the freehand draw operation, as well as the vertices of each point along the way:

5.1.3 MANUALLY BUILDING A ROI: THE "ROI DESCRIPTOR"

It is often beneficial to programmatically define ROIs, especially if there are several objects of interest that are always in the same place within a series of

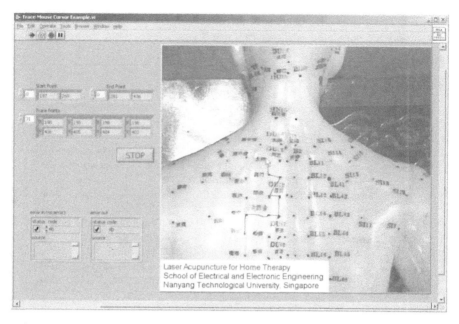

FIGURE 5.14 Trace mouse cursor example.

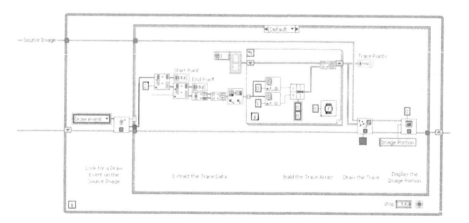

FIGURE 5.15 Trace mouse cursor example — wiring diagram.

images (e.g., parts on a production line). Setting the upper left and lower right corners is not enough to fully define a ROI, as it contains several other attributes. The ROI itself is carried around LabVIEW diagrams as a special cluster called the *ROI Descriptor* (Figure 5.16; Table 5.2).

Figure 5.17 is an example of a dynamically defined rectangular ROI. As the user changes the Manual Rect ROI values, a new ROI Descriptor is calculated with an internal ID. The image data within the ROI is then extracted and displayed as a new

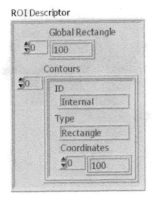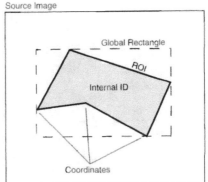

FIGURE 5.16 The ROI descriptor.

TABLE 5.2

Item	Description
Global rectangle	The bounding box of the ROI — a rectangle whose bottom side is parallel to the bottom of the source image and encloses the complete ROI, irrespective of the ROI's shape.
Contours	Accurately define the ROI with the following components: *ID:* Specifies whether the ROI is considered to be within or without the defined shape. If the ID is external, the ROI is the complete image outside of the ROI coordinates and is not bound by the global rectangle. *Type:* Specifies the contour shape (Point, Line, Rectangle, Oval, Polygon, Free, Broken Line, or Free Hand Line). *Coordinates:* Specifies the list of contour points (e.g., the top left and bottom right points for a rectangle)

---

image (the example in Figure 5.18 uses the global rectangle as the coordinates, as the type is a rectangle).

One often-useful application of dynamic ROI definition is an *inspection routine*, which allows the user to pan and zoom a source image to locate features (Figure 5.19).

Although the wiring diagram in Figure 5.20 may look complicated, it is just an extension of previous examples. An event handler waits on the user pressing one of the inspection navigation buttons on the front panel and adjusts either the window zoom or ROI descriptor to suit. The new values are then fed into IMAQ WindZoom and IMAQ WindSetROI and the inspection window is redrawn to reflect the changes.

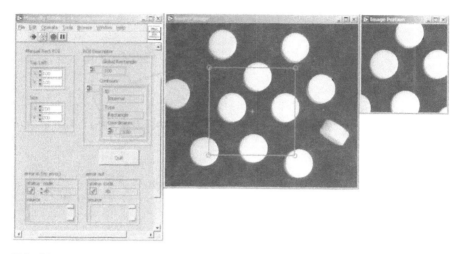

FIGURE 5.17 Manually building a rectangular ROI.

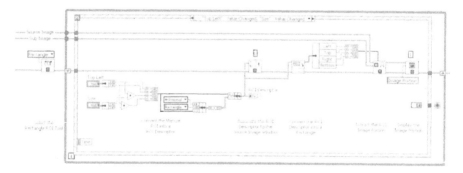

FIGURE 5.18 Manually building a rectangular ROI — wiring diagram.

FIGURE 5.19 Manual rectangular ROI application.

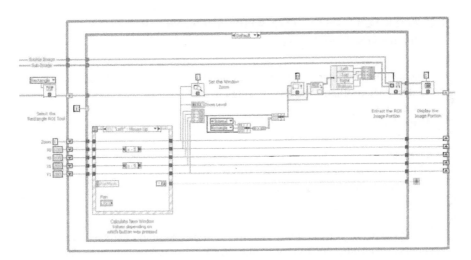

FIGURE 5.20 Manual rectangular ROI application — wiring diagram.

5.2 USER SOLUTION: DYNAMIC MICROSCOPY IN BRAIN RESEARCH

Michael Noll-Hussong and Arthur Konnerth
Ludwig-Maximilians-University Munich
Institute of Physiology
Pettenkofer Str. 12
D - 80336 Muenchen
Germany
Tel +49 (0)89 5996 571/Fax +49 (0)89 5996 512
Web site: http://www.neural.de
e-mail: minohu@gmx.net

The study of nerve cells is today one of the most exciting challenges in research, especially because the biology of elementary processes like "learning" and "memory" are not yet satisfactorily understood. Major progress toward a deeper understanding of the brain's functional mechanisms has been achieved with the development of imaging techniques, both macroscopic (e.g., employing positron-emission tomography) and microscopic (e.g., confocal and multiphoton microscopy). Latter mode-locked laser-based microscopy techniques have made it possible to visualize, with a high temporal and spatial resolution, the dynamics of life processes at the subcellular level, thus yielding new insights into the function of our brain on the cellular level. Of particular importance in this respect is the study of the changes in the intracellular calcium concentration in nerve cells, given calcium's involvement in the release of neurotransmitters and in memory processes. Specific fluorescent indicator dyes can be introduced into living cells, which emit light of varying wavelengths and intensities depending on the intracellular calcium concentration, thus making life processes accessible to direct visualization.

Our laboratory employs a range of microscopy systems of differing quality and functional characteristics. They are mostly integrated in electrophysiological workbenches, so that in the course of an experiment both imaging and conventional electrophysiological characterization of nerve cells are possible.

Addressing some fundamental questions of the physiology of our nervous system continually makes new demands on both the hardware (in its broadest meaning) and the software used for investigations. The limitations encountered in our earlier studies made it very soon apparent that software development is an integral component of our research. We soon reached the conclusion that would have to construct modular software appropriate to our own needs and after an intensive search we came across the programming language LabVIEW, which, as far as we were concerned, differed positively from its competitors in both its orientation towards measurement techniques and its integration of hardware and software components. Above all, the possibility of developing our own, checkable program codes and the capacity to adapt flexibly to new demands spoke strongly in favor of this development environment.

As examples of our developments, we introduced program components for acquisition (*FastRecord* module) and analysis (*FastAnalysis* module) of image data obtained in the course of our brain research using laser microscopy. The entire software suite was developed using LabVIEW 6i with IMAQ-Vision 6.0.5 and the hardware employed included image capture cards (NI PCI-1408 and NI PCI-1409) and a DAQ card (NI PCI-6111E). The image acquisition module allows lossless recording of video data to a hard disk in real time, with analysis functions being available simultaneously. A subsequent part of the program allows more detailed analysis of the experimental data and prepares them for final analysis in appropriate programs (*MS Excel*, *Wavemetric Igor*™). Further modules of our software allow networking with other data sources (e.g., electrophysiological parameters, temperature, etc.) and the integration of control logic (e.g., controlling microscope motors). Other modules include procedures for image analysis in static fluorescence microscopy with an *auto-ROI* function and an *alignment* procedure based on structure recognition for *antidrift* correction in prolonged sequences of images.

5.2.1 Program Description

The modules described were integrated in the program on the basis of a menu-driven queued state machine and thus could be executed directly via the GUI, or as individual menu items, whereby a uniform variable space definition ensures data exchange between the individual program components.

5.2.2 Recording Video Sequences

For high spatial and temporal resolution recording of the dynamic changes in the fluorescence characteristics of indicator dyes, a *recording* module (FastRecord) was developed to achieve the following:

- Lossless, real-time display of the biological object under investigation on the computer monitor. In this case the (analogue) image delivered by the microscope is processed directly by the computer program.
- Placement of ROIs over interesting regions of the image, interleaved with lossless, real-time display of brightness-versus-time courses.
- Direct and lossless storage of image data on the hard disk.
- Microscope control via the serial interface (e.g., laser shutter control, control of the z-drive).
- Communication with the electrophysiological module, for instance via trigger connections (circuits) and networking of and with measurement data from other sources.
- Simplicity and standardization of program operation, particularly with respect to appropriateness for the task in hand, capacity for self-description and conformity with expectation (in accordance with EN ISO 9241).
- Modular software concept with the option of expansion of the program's functionality at any time.

The procedure in the course of an experiment begins with the biological object of interest (a single cell in a vital brain section) is displayed as a video image on the PC monitor.

ROIs of any shape can be located at specific positions in the image as desired. The mean brightness value of these ROIs can be displayed simultaneously beside the live image as brightness-versus-time courses on the monitor. If required, pressing a key stores image sequences, theoretically of any desired length, without loss and in real time. The only limit to the recording duration is the storage capacity of the hard disk. The possibility for direct-to-disk storage is a major advance, since this circumvents the limitations inherent in the transient and nonpermanent storage in the expensive RAM and thus allows direct and permanent storage of digital films using the economic and theoretically unlimited capacity of the hard disk. In addition, it is possible to communicate with the various peripheral experimental instruments (laser shutter, microscope z-drive, etc.) and trigger connections allow data exchange with a so-called "patch-clamp" amplifier, so that the start of an electrophysiological experiment can initiate the recording of an image sequence and vice versa.

5.2.2.1 Analysis of Video Sequences

The primary aim of the FastRecord module is the comfortable and comprehensive analysis of image sequences of various formats (AVI, Bio-Rad™-PIC, BIN) obtained using FastRecord or other recording systems. The main focus thereby is the

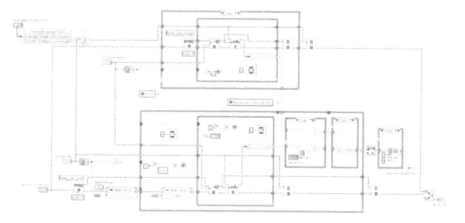

FIGURE 5.21 Calculate brightness vs. time.

capacity to produce (after definition of the appropriate ROIs) brightness-versus-time courses and to store the data in a format that can be read by conventional spreadsheet programs (Figure 5.21).

The procedure for the calculation of the brightness-versus-time courses in calcium-imaging experiments is as follows: the mean brightness value (F) of a ROI is calculated dynamically by summing the brightness values of all pixels within the ROI, dividing the sum by the number of pixels and rounding off mathematically to the nearest integer. The control brightness value (F_0) for the ROI is calculated by summing the individual images in the video sequence up to the point at which the evoked (electrophysiological) event commences, dividing the sum by the number of individual images and then calculating the mean brightness for each ROI. After correcting (by subtraction) both F and F_0 for the background brightness (background fluorescence), the quotient $\Delta F/F_0$ is obtained and displayed as the result of the analysis in the corresponding curves, which can also be stored in a conventional spread-sheet format (e.g., Excel, Wavemetrics Igor) for final analysis, if such is necessary.

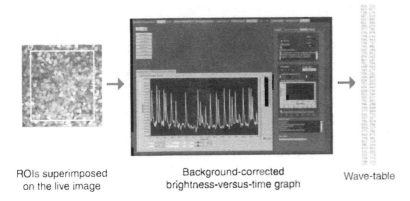

ROIs superimposed   Background-corrected   Wave-table
on the live image   brightness-versus-time graph

FIGURE 5.22 The FastRecord analysis module.

5.2.3 SUMMARY

The program modules FastRecord and FastAnalysis represent a complete solution for the recording and analysis of (analogue) fluorescence data with high temporal and spatial resolution. FastRecord (Figure 5.22) offers PC-based, real-time and lossless recoding of image data, with the possibility of the simultaneous monitor display of the object under investigation as well as the graphical representation of the live fluorescence values. The data obtained with FastRecord can be analyzed and prepared for presentation with FastAnalysis. The internal coordination and modularity of the program assures simple and intuitive use, with the possibility of expansion and integration of further functions at any time. The use of the widespread PC platform makes this program an economic alternative for carrying out imaging experiments, the qualitative (functional appropriateness) and quantitative (processing speed) characteristics are similar to those of commercially available imaging systems. As our experience has shown, the modularity of the above programs allows for flexible integration in larger applications that, as LabVIEW-based complete software, accompany the entire course of an experiment: managing, controlling, measuring and analyzing.

5.3 CONNECTIVITY

The nature of image processing alters a PUI's (pixel under investigation) value with respect to its surrounding pixel values and perhaps the PUI itself. This process is called *connectivity* — it determines which pixels to use in the computation of new values. Most Vision Toolkit image processing and morphological routines are based on 4, 6 or 8 pixel connectivity. Consider an image with its pixels referenced as their position within the image array:

00	01	02	03	04	05	06
10	11	12	13	14	15	16
20	21	22	23	24	25	26
30	31	32	33	34	35	36
40	41	42	43	44	45	46

Four-pixel connectivity concerns the image processing routine with the cardinal (i.e., North, East, South and West) pixels, so if the PUI is 23, then the connectivity partners are 13, 22, 24 and 33:

00	01	02	03	04	05	06
10	11	12	**13**	14	15	16
20	21	**22**	**23**	**24**	25	26
30	31	32	**33**	34	35	36
40	41	42	43	44	45	46

Six-pixel connectivity represents a hexagonal pixel layout, which can be simulated on a rectangular grid by shifting each second row by half a pixel:

00	01	02	03	04	05	06
10	11	12	13	14	15	16
20	21	22	23	24	25	26
30	31	32	33	34	35	36
40	41	42	43	44	45	46

$\Rightarrow$

00	01	02	03	04	05	06	
	10	11	12	13	14	15	16
20	21	22	23	24	25	26	
	30	31	32	33	34	35	36
40	41	42	43	44	45	46	

If PUI were 23, its 6 pixel connectivity partners would be 12, 13, 22, 24, 32 and 33:

00	01	02	03	04	05	06	
	10	11	12	13	14	15	16
20	21	22	23	24	25	26	
	30	31	32	33	34	35	36
40	41	42	43	44	45	46	

Eight-pixel connectivity concerns the image processing routine with the cardinal and subcardinal (i.e., North East, South East, South West and North West) pixels, so if the PUI is 23, then the connectivity partners are 12, 13, 14, 22, 24, 32, 33 and 34:

00	01	02	03	04	05	06
10	11	12	13	14	15	16
20	21	22	23	24	25	26
30	31	32	33	34	35	36
40	41	42	43	44	45	46

5.4 BASIC OPERATORS

Basic operators execute by performing a pixel-by-pixel transformation on the intensities throughout the source image(s) and function in two modes: with two source images, or with one source image and a constant operator.

5.4.1 ADD (COMBINE)

Adding two source images is performed as simple matrix addition:

$$Dst(x, y) = SrcA(x, y) + SrcB(x, y)$$

A1	A2	A3
A4	A5	A6
A7	A8	A9

+

B1	B2	B3
B4	B5	B6
B7	B8	B9

=

A1+B1	A2+B2	A3+B3
A4+B4	A5+B5	A6+B6
A7+B7	A8+B8	A9+B9

Source Image A   Source Image B   Combined Images

Therefore, an example 3×3 image combination would be implemented as follows:

122	127	125
190	120	123
115	178	160

+

122	0	111
187	130	113
117	180	167

=

244	127	236
377	250	236
232	358	327

Source Image A   Source Image B   Combined Images

As intensity values greater than 255 are forbidden in an 8-bit image (only values between 0 and 255 inclusive are permitted), any combined intensity values of 256 or greater (shaded cells in the previous example) are clipped to 255 (white). Therefore, the true (distorted) output of the combination routine for this example is:

122	127	125
190	120	123
115	178	160

+

122	0	111
187	130	113
117	180	167

=

244	127	236
255	250	236
232	255	255

Source Image A   Source Image B   Combined Images

As you can see, the combined image's intensity values are the added components and are therefore either larger than (hence brighter) or equal to that of the original pixels. The example in Figure 5.23 was implemented with the code shown in Figure 5.24, using IMAQ Add. IMAQ Add also allows for constant values to be added

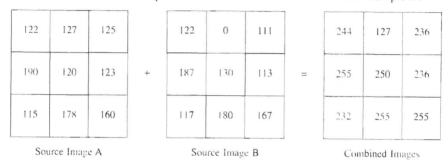

(a)   (b)   (c)

FIGURE 5.23 (a) Source image A (b) Source image B (c) Combined images

to all image pixels — the constant input can be either positive or negative (Figure 5.25). As you might expect, using a constant of 100, adds 100 intensity units to each pixel of the source image (Figure 5.26).

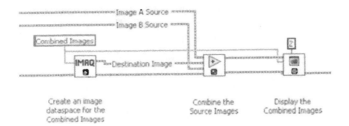

FIGURE 5.24 Add images — wiring diagram.

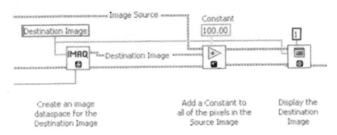

FIGURE 5.25 Add constant to image — wiring diagram.

FIGURE 5.26 Add constant to image.

5.4.2 SUBTRACT (DIFFERENCE)

Subtraction is also performed as a simple matrix operation:

$$Dst(x,y) = SrcA(x,y) - SrcB(x,y)$$

A1	A2	A3
A4	A5	A6
A7	A8	A9

Source Image A

B1	B2	B3
B4	B5	B6
B7	B8	B9

Source Image B

A1-B1	A2-B2	A3-B3
A4-B4	A5-B5	A6-B6
A7-B7	A8-B8	A9-B9

Combined Images

Therefore, an example 3×3 image combination would be implemented as follows:

122	127	125
190	120	123
115	178	160

Source Image A

$-$

122	123	111
187	130	113
117	180	167

Source Image B

$=$

0	4	14
3	-10	10
2	-2	-7

Combined Images

As intensity values less than 0 are forbidden in a 256 level image (only values between 0 and 255 inclusive are permitted), any combined intensity values of 0 or less (shaded cells in the previous example) are clipped to 0 (black). Therefore, the true (distorted) output of the combination routine for this example is:

122	127	125
190	120	123
115	178	160

Source Image A

$-$

122	123	111
187	130	113
117	180	167

Source Image B

$=$

0	4	14
3	0	10
2	0	0

Combined Images

The subtracted image's intensity values are the subtracted components and are therefore either smaller than (hence darker) or equal to that of the original pixels. The example in Figure 5.27 shows an image of a digital clock, from which an intensity-graduated box is subtracted. As you can see, the corresponding portions of the resulting image located where the *gray boxes* image is dark are relatively unchanged, as the subtracting factor is dark (therefore close to zero). Conversely, where the *gray boxes* image is bright, the pixel intensity values are relatively high and therefore the resulting image's pixels are low. The example in Figure 5.28 shows the opposite scenario — the clock image is subtracted from the gray boxes image. The examples in Figure 5.28 were implemented using IMAQ Subtract, shown in Figure 5.29.

IMAQ Subtract also allows for constant values to be subtracted from all image pixels (Figure 5.30). As you might expect, using a constant of 100 subtracts 100 intensity units from each pixel of the source image (Figure 5.31).

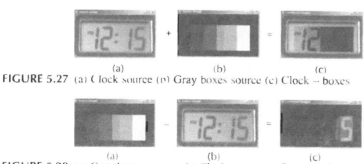

(a)   (b)   (c)

FIGURE 5.27 (a) Clock source (b) Gray boxes source (c) Clock – boxes

(a)   (b)   (c)

FIGURE 5.28 (a) Gray boxes source (b) Clock source (c) Boxes – clock

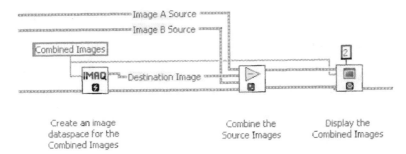

Create an image
dataspace for the
Combined Images

Combine the
Source Images

Display the
Combined Images

FIGURE 5.29 Subtract images — wiring diagram.

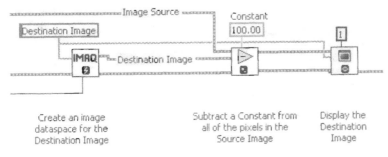

Create an image
dataspace for the
Destination Image

Subtract a Constant from
all of the pixels in the
Source Image

Display the
Destination
Image

FIGURE 5.30 Subtract constant from image — wiring diagram.

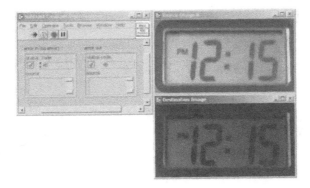

FIGURE 5.31 Subtract constant from image.

5.4.3 OTHER BASIC OPERATORS

Operator	Description

IMAQ Multiply

Multiplies the pixels of two source images, or one source image with a constant. Be careful to match the source and destination images so that pixel intensity clipping does not occur (e.g., decrease the intensity range of the source images, or increase the destination image).

$$Dst(x, y) = \frac{SrcA(x, y)}{SrcB(x, y)}$$

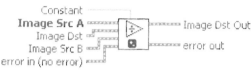

IMAQ Divide

Divides the pixels of the first source image with the pixels of the second source image or a constant.

$$Dst(x, y) = \frac{Const\{SrcA(x, y)\}}{SrcB(x, y)}$$

IMAQ MulDiv

Computes a ratio between the two source images. Each pixel in Source Image A is multiplied by a constant and then divided by its equivalent Source Image B pixel. Unlike other basic operators, a temporary variable is used to perform this operation; so information is not lost due to compression

$$Dst(x, y) = \{SrcA(x, y) - SrcB(x, y)\} \times$$
$$\left\lfloor \frac{SrcA(x, y)}{SrcB(x, y)} \right\rfloor$$

IMAQ Modulo

Executes modulo division (remainder) of one image by another or an image by a constant

FIGURE 5.32

5.4.4 AVERAGING

Image averaging is a well-tested method of minimizing acquired random image noise, hence increasing the *signal-to-noise ratio* (SNR). A common error when attempting image averaging is to add two images and then divide them by two, hoping for an image average (Figure 5.33). Although this technique will work when the added pixel values are not clipped to the maximum intensity, if the addition does equal more than the maximum, compression will occur, as shown in Figure 5.34.

A better method is to scale down each of the source images before the addition (Figure 5.35) however, this reduces the dynamic range if rounding occurs, which achieves the result shown in Figure 5.36.

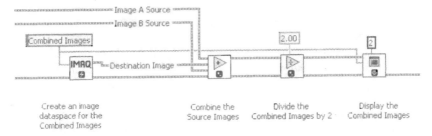

FIGURE 5.33 Average images: The wrong way — wiring diagram.

FIGURE 5.34 (a) Source image A. (b) Source image B (c) The wrong way.

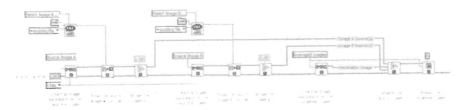

FIGURE 5.35 Average images: A better way — wiring diagram.

FIGURE 5.36 (a) Source image A (b) Source image B (c) A Better Way

5.5 USER SOLUTION: IMAGE AVERAGING WITH LABVIEW

Don J. Roth
Senior Research Engineer
NASA Glenn Research Center
Cleveland, Ohio
e-mail: don.j.roth@grc.nasa.gov

Many data acquisition processes are besieged by noise, including those that are image-based.[1] *Signal averaging* is a simple and popular technique to reduce noise or increase the *signal-to-noise ratio* (SNR). SNR is proportional to the square root of the number of averages. Based on the circular buffer technique, the routines I present allow the running image average of the most recent x number of images, with the number being continuously adjustable in real-time, without disruption or discontinuity of the averaging computation. The advantage of this is that you can manage the trade-off between noise reduction and the amount of processing in the time-domain during the test, if so desired. In other words, you can tune to, or "dial in," the level of SNR desired while the experiment is running and actually see the *in-situ* effect on the image (Figure 5.37).

When dealing with continuous single shot measurements, a one-dimensional array is needed to hold the collection of point measurements. When dealing with waveforms, a two-dimensional array is needed to hold the collection of one-dimensional waveforms. When dealing with images, a three-dimensional array is needed to hold the two-dimensional image frames. LabVIEW functions (such as `initialize array`, `array subset` and `insert into array`) need to be expanded to deal with the different array sizes in each of the averaging schemes, as shown below. This is achieved by right clicking on the appropriate input terminal of the function and selecting "add dimension" (Figure 5.38).

Image averaging was performed by indexing out the columns of the three-dimensional array and then using the LabVIEW Mean function (from the *Mathematics, Probability and Statistics* subfunction palette) on each column (one-dimensional array) within the three-dimensional array. To better visualize this, consider each image frame as a checkerboard of squares (pixels). Stack the checkerboards on top of each other. Now extract out the vertical columns at every location and calculate the mean of each one. The result is a two-dimensional array of mean values (or an image average) (Figure 5.39).

The National Instruments Developer Zone shows an example running image averager using IMAQ Vision functions[2] that operates somewhat differently than the one presented here. In the National Instruments example, after a specified number of images are averaged, the averaged image becomes the first image for the new set of images to be averaged, whereas the routines presented here allow the running image average of the most recent x number of images.

[1] Roth, Don J., *LabVIEW Technical Resource*, Vol. 8, No. 4, pp. 12–14.
[2] NI Developer Zone (http://zone.ni.com/) Example: Real-time Image Averaging.vi.

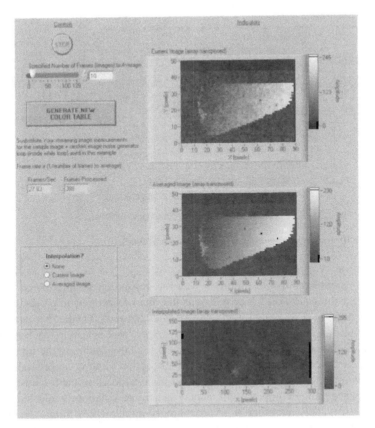

FIGURE 5.37 Image averaging alternatives — Front panel example.

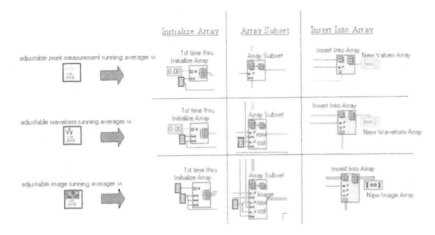

FIGURE 5.38 Image averaging alternatives — comparison with other averagers.

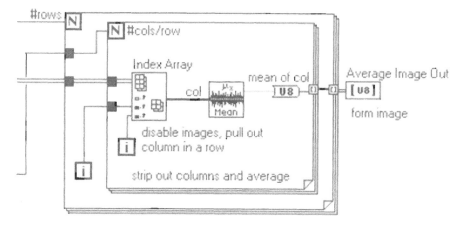

FIGURE 5.39 Image averaging alternatives — how it works.

In the example shown in Figure 5.37, a low-resolution sample image (89 × 44 pixels) plus simulated noise is provided as the input to demonstrate these routines. An unsigned 8-bit (U8) format with images scaled between 0 and 255 is used. Array transposition for the version using intensity graphs was necessary. A two-dimensional interpolation routine[1] has been integrated to allow enlarging the image while retaining the smoothing effect. The size of the active plotting area is adjusted in terms of width and height in pixels based on the interpolation factor chosen (interpolation factor numeric control will display when interpolation is selected in radio button control). The interpolation routine is only implemented in this example for the intensity graph container but can easily be integrated into the examples using IMAQ and Picture Control display containers.

For practical use of these routines, one would substitute NI-IMAQ routines that perform continuous live image acquisitions from a CCD or other type of camera. The frame rate one obtains will be inversely proportional to the number of frames being averaged (as well as the physical size/number of pixels of the frames and size of intensity graph indicator). On a Compaq 1.5 GHz Xeon single processor-based system, I was able to process about 50 fps with 10 frames averaged and a 1-ms delay in the while loop of the block diagram. Keep in mind a two-dimensional array of noise is calculated within each loop iteration while for an actual machine vision situation, frame grabbing would be occurring instead and this will affect the true frame rate achieved. A color table control routine is also included, allowing the user to modify the color table during acquisition, although a gray scale color table is provided as the default.

---

[1] NI Developer Zone Example: 2D Interpolation demo.llb.

5.6   OTHER TOOLS

Most of the other tools in the Image Processing palette of the Vision Toolkit speak for themselves, but there are a few that are not completely intuitive.

5.6.1   Symmetry (Mirroring an Image)

IMAQ Symmetry transforms an image to mirror along one or more axes (Figure 5.40). Using IMAQ Symmetry is quite simple — all you need to define is which axes to use (Figure 5.41).

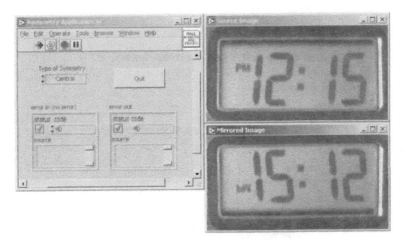

FIGURE 5.40 Symmetry application.

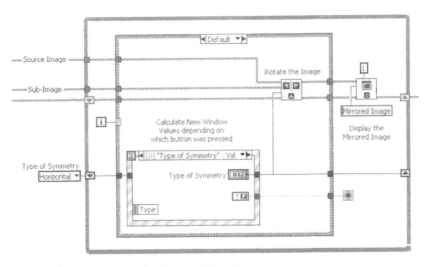

FIGURE 5.41 Symmetry application — wiring diagram.

The available types of symmetry are:

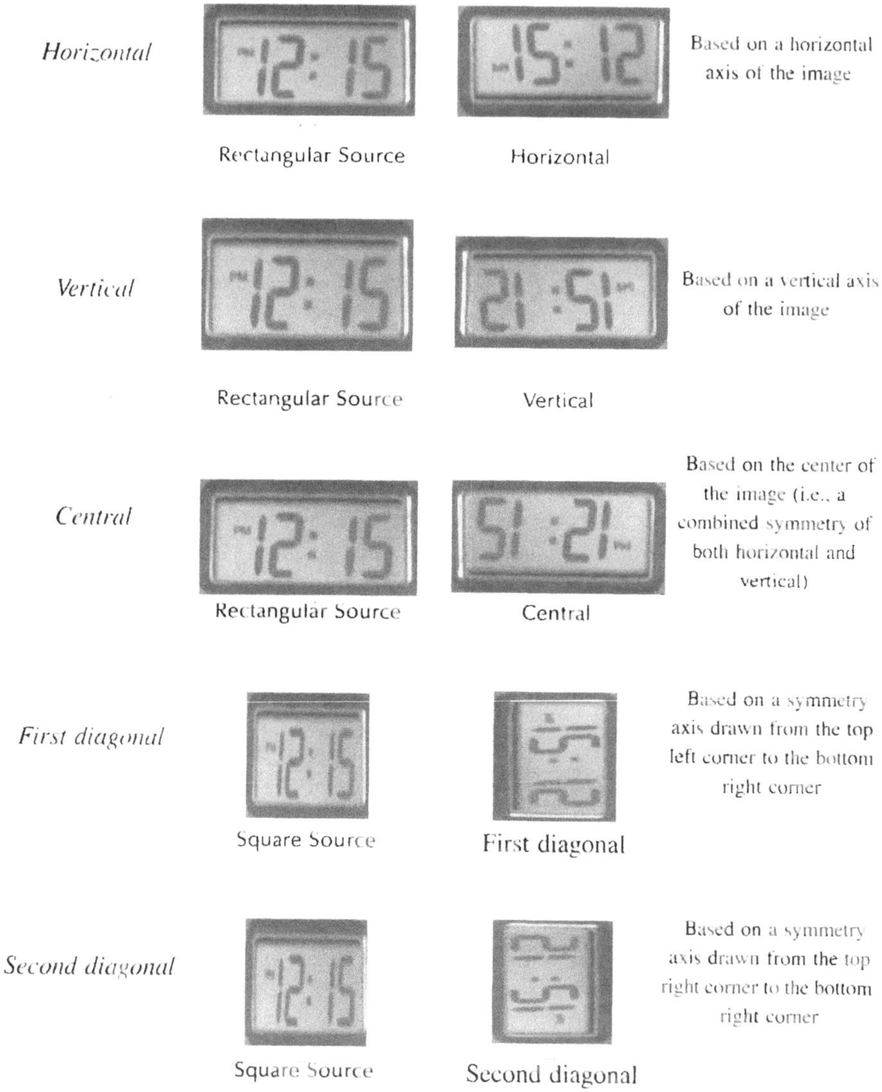

Horizontal	Rectangular Source	Horizontal	Based on a horizontal axis of the image
Vertical	Rectangular Source	Vertical	Based on a vertical axis of the image
Central	Rectangular Source	Central	Based on the center of the image (i.e., a combined symmetry of both horizontal and vertical)
First diagonal	Square Source	First diagonal	Based on a symmetry axis drawn from the top left corner to the bottom right corner
Second diagonal	Square Source	Second diagonal	Based on a symmetry axis drawn from the top right corner to the bottom right corner

FIGURE 5.42

5.6.2 ROTATE

IMAQ Rotate allows you to rotate an image with single point float accuracy (Figure 5.43). This example (Figure 5.44) was achieved using a modified version of Manual Rectangular ROI Application.vi, discussed earlier.

As you can see, when image data is rotated, the image window itself does not rotate, leaving "blank" areas around the newly rotated section. The image data for these areas is undefined, so IMAQ Rotate allows you to specify a *replace value* to paint them with. The replace value is specified in intensity units between 0 and 255.

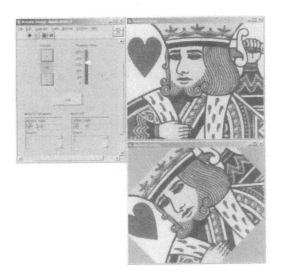

FIGURE 5.43 (a) Rectangular source. (b) Horizontal.

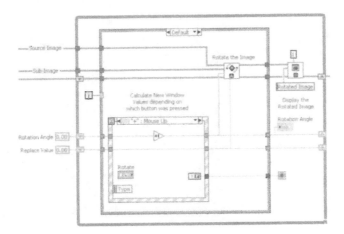

FIGURE 5.44 Rotate image application — wiring diagram.

IMAQ Rotate also allows you to specify whether the pixel interpolation used to complete the rotation be either zero-order or bilinear. As Bruce Ammons from Ammons Engineering recently pointed out on the National Instruments Developer Zone Discussion Forum:

> [The Vision Toolkit] interpolation is a two-step process. When calculating the intensity of each new pixel, the first step is to determine the coordinates of the pixel within the original image. The coordinates are usually between existing pixels. The second step is [to] do the interpolation. Zero order is just picking whichever pixel is closest to the coordinates. Bi-linear interpolation uses a weighting function that combines the intensities of the 2 × 2 grid of pixels surrounding the coordinates. It is just like linear interpolation between two points, but it is done in 3D.

5.6.3 Unwrap

Often images contain information that is wrapped in a circular pattern, like text printed on the inner ring of a CD, or a produce jar lid (Figure 5.45). The inset window contains the unwrapped image. First, you need to define the annulus arc within the image that is to be unwrapped, which can be defined manually or dynamically by the user drawing out an annulus arc ROI. Once the annulus arc has been defined, it is then fed into IMAQ Unwrap, along with the original image (Figure 5.46). IMAQ Unwrap then distorts the image data within the annulus arc into a rectangle, so it is inevitable that information will be changed during this transformation — the new pixel values are determined by either zero-order or bilinear interpolation, as described in the Rotate section.

FIGURE 5.45 Unwrap.

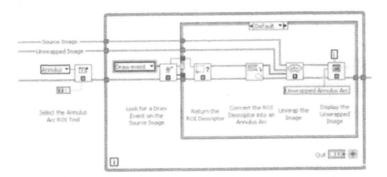

FIGURE 5.46 Unwrap — wiring diagram.

5.6.4 3D View

A grayscale image has three dimensions — each pixel on a two-dimensional spatial plane has an intensity component, hence the third dimension. An often-useful method of displaying these intensity maps is by using the Vision Toolkit's 3D View, where each pixel's intensity value is represented spatially (Figure 5.47). As you can see, the three-dimensional representation is also artificially shaded. This technique is achieved using IMAQ 3D View (Figure 5.48). IMAQ 3D View is also useful to plot non-Vision two-dimensional arrays by first converting them to IMAQ images (Figure 5.49).

FIGURE 5.47 3D view.

The example in Figure 5.50 shows a two-dimensional array dataset plotted first as a image (in a similar fashion to an intensity graph) and then as a 3D View image.

5.6.5 Thresholding

Thresholding enables you to select ranges (often referred to as the *threshold interval*) of pixel intensity values in grayscale and color images, effectively compressing the values outside of the range to their respective extremes.

$$I_{New} = \begin{cases} 0 & I_{Old} < Range_{Low} \\ I_{Old} & Range_{Low} \leq I_{Old} \leq Range_{High} \\ 255 & I_{Old} > Range_{High} \end{cases}$$

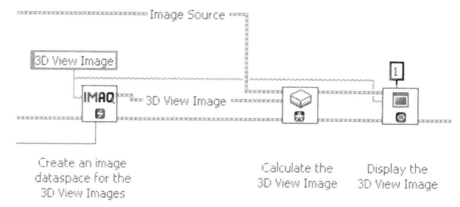

FIGURE 5.48 3D view — wiring diagram.

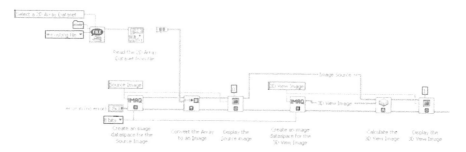

FIGURE 5.49 3D view: plot two-dimensional array — wiring diagram.

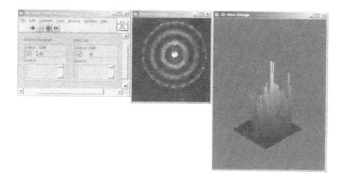

FIGURE 5.50 3D view — plot two-dimensional array.

Consider a grayscale image:

125	122	111	110	107
130	145	155	134	123
131	156	159	141	136
135	177	180	149	142
150	202	213	170	160

Perhaps you are interested only in intensity values that fall within a particular range, e.g., 140 and 175 inclusive. Performing a threshold operation on the previous image would yield the following image:

0	0	0	0	0
0	145	155	134	0
0	156	159	141	0
0	255	255	149	142
150	255	255	170	160

You can also perform a binary threshold by suppressing the intensities outside of the threshold interval to zero and highlighting those within the interval by setting their values to the maximum intensity values.

$$I_{New} = \begin{cases} 0 & I_{Old} < Range_{Low}, I_{Old} > Range_{High} \\ 255 & Range_{Low} \leq I_{Old} \leq Range_{High} \end{cases}$$

Consider the previous image again. Performing a binary threshold using the same interval values of 140 and 175 yields:

0	0	0	0	0
0	255	255	255	0
0	255	255	255	0
0	0	0	255	255
255	0	0	255	255

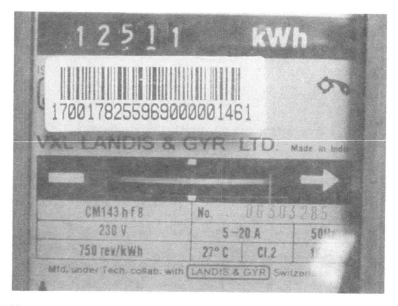

FIGURE 5.51 Power meter with barcode label.

For example, Figure 5.51 contains a barcode label. The label is relatively bright when compared with the rest of the image, so using a nonbinary threshold routine that sets pixels with intensities higher than a value that is just lower than the label will achieve the result shown in Figure 5.52.

When used in binary mode, IMAQ Threshold also has a *replace value* input, so you can specify the intensity value to be used when the PUI's value falls within the threshold interval (Figure 5.53). For example, setting a custom replace value of 100 to the previous matrix image results in:

0	0	0	0	0
0	100	100	100	0
0	100	100	100	0
0	0	0	100	100
100	0	0	100	100

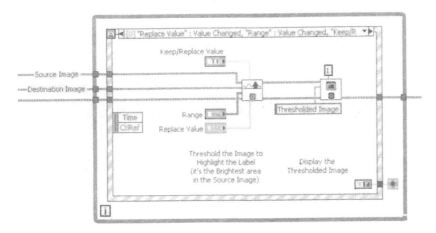

FIGURE 5.52 Threshold example.

Applying the same replace value of 100 to the barcode label image above yields the result in Figure 5.54.

FIGURE 5.53 Threshold example — wiring diagram.

FIGURE 5.54 Threshold example — custom replace value.

5.6.6 EQUALIZATION

Equalization is similar to thresholding an image, except that it also spreads the threshold interval intensities across the complete range permitted by the image type, redistributing the pixel values to linearize the accumulated histogram. Equalization increases the contrast of a particular range of intensities.

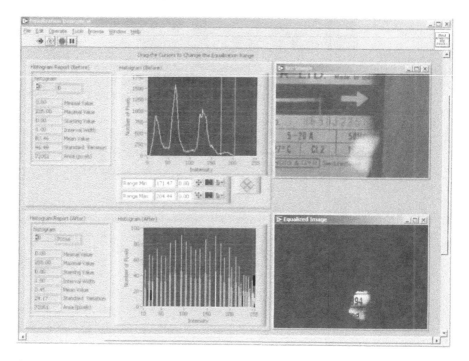

FIGURE 5.55 Equalization example.

For example, the image below contains a glare spot, where incident light has reflected from a glossy surface, creating a blind spot. Equalization can be used to enhance the image, effectively stretching the *equalization range* to cover the complete intensity range, making it much easier to read the underlying text. The equalization example in Figure 5.55 was achieved using the code shown in Figure 5.56.

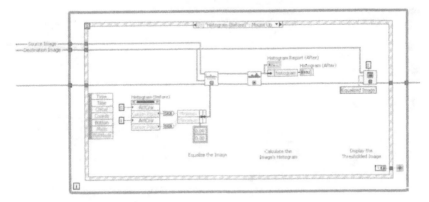

FIGURE 5.56 Equalization example — wiring diagram.

TABLE 5.3

Item	Description
Minimum	The minimum value is the smallest intensity to use during equalization. After the equalization is complete, all pixel values that are less than or equal to the minimum in the original image are set to 0 for an 8-bit image. In 16-bit and floating-point images, these pixel values are set to the smallest pixel value found in the original image.
Maximum	The maximum value is the highest intensity to use during equalization. After the equalization is complete, all pixel values that are greater than or equal to the maximum in the original image are set to 255 for an 8-bit image. In 16-bit and floating-point images, these pixel values are set to the largest pixel value found in the original image.

The equalization range is a cluster containing the elements described in Table 5.3. Equalization can also be used to remove background data when an object's intensities are very different from that of the background (Figure 5.57). Notice that the background is completely removed and the contrast of the leaves has been increased.

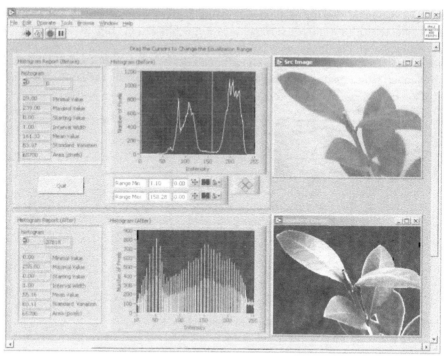

FIGURE 5.57 Equalization example — removing a background.

5.7   USER SOLUTION: QUICKTIME FOR LABVIEW

Christophe Salzmann
Swiss Federal Institute of Technology
Lausanne, Switzerland
e-mail: christophe.salzmann@epfl.ch
Web site: http://iawww.epfl.ch

QuickTime is a free multimedia package from Apple (a fee-based Pro version is available, but not required to use the LabVIEW interface). It allows users to perform a wide variety of operations on images, movies and sounds, of which a subset has been provided to LabVIEW with the help of a CIN. As both QuickTime and LabVIEW are platform independent, code written to exploit their strengths is also cross-platform.

QuickTime can be seen as a set of APIs that supports a wide variety of multimedia tasks, implemented as a set of extensions under MacOS and DLLs under Windows. Over the years, QuickTime has been extended to support many of the current formats for compressing, transporting and storing images, videos and sounds and also handles 3D and VR objects, cubic panorama, MIDI music, etc. Many of the QuickTime functions are hardware accelerated, leveraging the graphics processor as opposed to using the CPU for operations. Hardware acceleration also permits applications to benefit from new hardware without requiring recompilation of the source code.

5.7.1   QUICKTIME INTEGRATION TO LABVIEW

The output of many QuickTime operations results in an image displayed in the user interface — the display of a given frame of a movie, for example. There are currently two cross-platform methods of displaying images in LabVIEW: the use of an external window and the use of the picture toolkit (although often popular, ActiveX solutions are not cross-platform and therefore are not considered in this text). The external window solution is similar to the IMAQ toolkit external window. The management of these external windows is not trivial and using the LabVIEW picture toolkit instead can alleviate many interfacing problems. A drawback of using the picture toolkit is the display speed of the LabVIEW picture indicator, but as the speed of the graphics card increases, this drawback tends not to be as noticed anymore. This solution also substantially simplifies the internal management of the library, as there is no .c code involved for the user interface, with native LabVIEW handling all aspects of it.

Great effort has been put in to this library, limiting the use of native QuickTime dialogs, replacing them with their LabVIEW equivalent, providing a greater flexibility and familiarity to the user. The user can ask the library to provide the needed information in order to build a custom dialog, listing the image formats can be requested and displaying the results in a dialog ring, for example.

5.7.1.1 QTLib: The QuickTime Library

In 1994, the ancestor of the QTLib existed for the *Connectix* (now *Logitech*) Quick-Cam, communicating through a serial link and displaying a grayscale grabbed image in a LabVIEW intensity graph. Since then, it has evolved to support more cameras and frame grabbers and the integration of the picture toolkit to the standard version of LabVIEW has permitted image display to be much faster. The current version of QTLib adds QuickTime functions to manage grabbed images or other front panel elements, with the ability to save them as single image or as a movie.

The QTLib is implemented using a *Fat-CIN*, a technique that permits cross-platform capabilities by embedding multiple compiled codes into one file, with the appropriate platform specific CIN automatically selected at run time using a dynamic call (see Cross Platform CIN by Christophe Salzmann, LabVIEW Technical Resource (LTR), Vol. 10, No. 1, 2002, at www.ltrpub.com for more information concerning Fat-CINs). There are six main usages for this library depicted by six examples VIs:

1. Reading single images
2. Writing single images
3. Reading movies (i.e., multiple images)
4. Writing movies
5. Video source frame grabbing
6. Image transformation

5.7.1.2 Single Image Reading and Writing

QT_ImgOpen.vi (Figure 5.58) attempts to find a QuickTime image importer that supports the files' encoding and will then decompress and return it in the Image Out cluster. Then QT_QTImageToLVPict.vi converts the image to a LabVIEW picture structure, appropriate to be drawn in a picture indicator. A warning is displayed if a suitable importer/decompressor was not found. The decompression operation can take some time and memory depending on the original image size and although QTLib will do its utmost to decompress the picture within the available memory, it may fail. If this does occur, the image should be decompressed in parts, although this option is not yet available.

The WriteImage example takes an array of bytes representing an image and writes them to the file specified by the user. As QT_ImgSave.vi expects the image an uncompressed 24-bit RGB image, you can convert a front panel element to this format using either PictToU8.vi or RefToU8.vi.

5.7.1.3 Reading Movies

The ReadMovie.vi (Figure 5.59) reads a selected movie frame by frame. Besides the native QuickTime movie format, various others are supported, including animated GIF, ShockWave Flash and MPEG. Although some movie files can also contain more than one track (an audio, text or second video track for example), currently QTLib does not support them — only the primary video track is used.

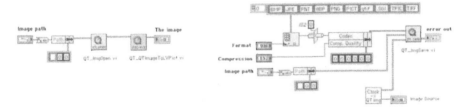

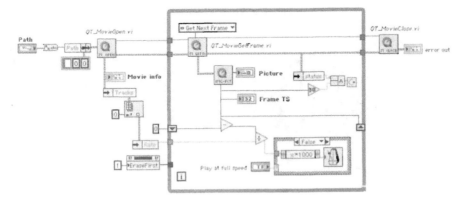

FIGURE 5.58 Using *ReadImage* and *WriteImage*.

FIGURE 5.59 Using *ReadMovie*.

Also, it might be necessary for the software to drop frames or to extend the time frames are displayed in order to play the movie in real-time, hence the *Play at full speed* button.

5.7.1.4 Writing Movies

Movie writing is similar to multiple sequential images writing (Figure 5.60). Currently, only spatial compression is supported, as using a format with temporal compression will trigger a new key frame for each frame. QT_MovieAddFrame.vi adds an image to the created movie and the frame duration (expressed in the movie time base ticks, not in seconds) is specified. Once all of the required frames are added, the movie is closed with QT_MovieClose.vi.

5.7.1.5 Video Grabbing (Figure 5.61)

Video grabbing was the first and only function of the initial version of QTLib and although the original version supported only one video input at a time, the current version supports up to 32 devices in parallel (Figure 5.61). QTLib has been successfully tested with USB, 1394 FireWire, serial and built-in digitizer boards. As described above, QTLib relies on video digitizer drivers to interface with image acquisition hardware and although most of them respect the QuickTime standard, some do so only partially or not at all, so you should confirm your hardware will support the standard fully before purchasing it. Under Windows, the original QTLib

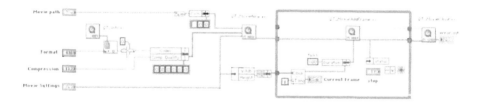

FIGURE 5.60 Using *WriteMovie*.

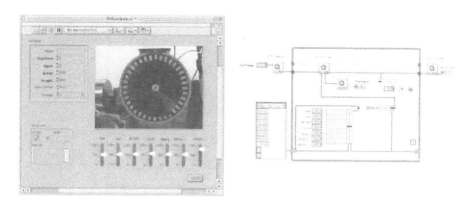

FIGURE 5.61 Video grabbing example.

supported Video For Windows, which has many limitations including the lack of concurrent inputs. The use of Tim Molteno's WinVDIG (see www.vdig.com for more information) alleviates this problem and permits the use of QuickTime standard APIs across platforms.

A frame will be grabbed at the specified size and returned as an image. QT_FGGrab.vi must be called as often has possible to keep up with the source's frame rate (each frame has an ID associated to it, so you can see if a frame has been missed). It is also possible to modify various settings of the video inputs, such as brightness and contrast, in real-time. Some video digitizer drivers have automatic settings that overcome user control, while others may not provide control for all the standard settings at all. The frame grabbing process occurs in parallel of other LabVIEW tasks and then must stopped by calling QT_FGClose.vi (this VI also releases all of the internal resources associated with the capture).

5.7.1.6 Image Transformation Functions

The image transformation function (Figure 5.62) applies one or more graphical operations (scale, rotation, skew, etc.) to a LabVIEW picture, or other LabVIEW front panel elements that can be accessed through the *Get Image* method. This includes picture indicators, graphs, charts, slides, strings, etc. For example, the TransformImage3.vi displays a 45° rotated thermometer. As picture indicators can return the mouse coordinates, it is possible to perform the reverse transformation to map the mouse click back in the original coordinate. Then the appropriate action, like changing a slider value, can be taken to provide a real interactive solution.

 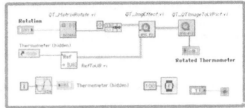

FIGURE 5.62 Using *TransformImage*.

5.7.1.7 Future Versions of QTLib

It is planned that QTLib will evolve in the future first to keep pace with new releases of QuickTime and LabVIEW and to add functions that could not be implemented in the current release of this library. The current version of the QTLib deals mostly with files stored on the hard drive, although a future release will access memory image maps as well. Also, more transformations and effects will be added to the image transformation VI.

5.7.1.8 For More Information

More detailed QuickTime information can be found at the Apple Web site (http://www.apple.com/quicktime) and an excellent introduction to QuickTime programming is *Discovering Quicktime: An Introduction for Windows and Macintosh Programmers*, by George Towner and Apple Computer, published by Morgan Kaufman, 1999.

5.8 FILTERS

Image filters are routines that either suppress or enhance data to meet a particular specified criterion. Typical filter uses include high contrast area enhancement, edge detection and smoothing.

The two general types of filters that you may come across are *linear* (often referred to as a *convolution*) and *nonlinear*. Linear filters are designed to recalculate the value of the PUI based on its original value and the values of those surrounding it. This convolution often weights the effect surrounding pixels have on the PUI, usually with respect to their distance from it (these weightings are reflected in the coefficients of the *convolution kernel*). An example convolution kernel is shown below:

0	0	0
0	1	0
0	0	0

The effect of this convolution kernel (hereafter referred to as the *filter*) is nothing. The recalculated value of the PUI is 1 time its original value, added to 0 times the values of those around it. A more useful example is:

0	1	0
1	1	1
0	1	0

This filter changes the value of the PUI to 1 time its original value, plus the sum of the North, East, South and West pixel values. Note: unless the intensity values of the used surrounding pixels are zero, the PUI's value will become higher (brighter). If the PUI or its surrounding pixels' intensities sum to a value higher than the maximum permitted for the image, intensity clipping will occur. A 256 level grayscale image example using the above filter could be:

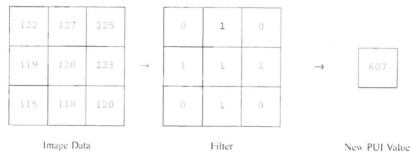

Image Data   Filter   New PUI Value

As an intensity value of 607 is forbidden in a 256 level image, the new PUI value will be clipped to 255 (white), which does not reflect the true nature of the filter. To avoid clipping, we can alter the previous kernel example to be a simple averager:

0	$1/_8$	0
$1/_8$	$1/_2$	$1/_8$
0	$1/_8$	0

This filter assumes that the value of the original PUI is the most important (giving it a weighting of $1/_2$) and those at the North, East, South and West each have equal importance of $1/_8$, so the same image example would be:

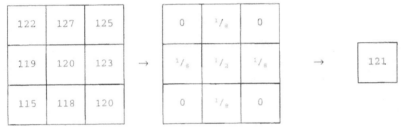

Image Data	Filter	New PUI Value

As all of the weightings of the filter add up to 1, there is no risk of the resulting PUI value exceeding the upper intensity limit of the pixel.

A 3×3 kernel with positive elements has been used in the examples above, but a grid of practically any size and shape, with elements positive, zero or negative values can be defined as required by your application. If a kernel contains only positive elements, such as in the examples above, the effect is that of smoothing "low pass" filter. Conversely, if the elements are mixed positive and negative, a sharpening "high pass" filter is achieved. The Vision Toolkit has a set of predefined kernels for Gaussian, gradient, Laplacian and smoothing filters in standard sizes of 3×3, 5×5 and 7×7, although you can easily define your own specialized filters of virtually any size.

5.8.1 Using Filters

You can use either the predefined filters built into the IMAQ Vision Toolkit, or create your own for specialized applications. To apply a filter, you simply select (or define) the appropriate kernel and then use IMAQ Convolute (Figure 5.63).

Connect your image source and kernel and IMAQ Convolute will do the rest. You may have also noticed the *Divider* (*kernel sum*) connector — it is a factor that is applied to the sum of the obtained products. Usually, you can leave this terminal unwired (it defaults to zero), but if it is not equal to 0, the elements internal to the matrix are summed and then divided by this factor. This is not the same as normalization, where the output is scaled to fall between defined values (0 and 1 for example), but is a raw operator.

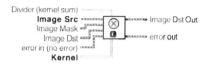

FIGURE 5.63 IMAQ convolute function.

5.8.2 PREDEFINED FILTERS

The Vision Toolkit has four built in predefined filter types: Gaussian, gradient, Laplacian and smoothing. You may discover that these types will cover most of the general purpose filtering that you will require. The IMAQ Vision online help contains a full list of the predefined filters available and their kernel definitions(Figure 5.64).

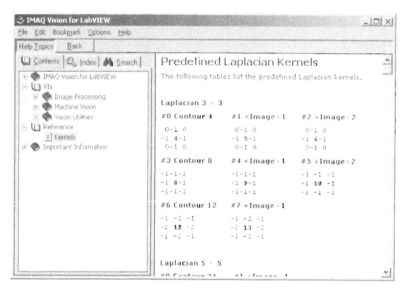

FIGURE 5.64 Online Help — predefined filters.

5.8.2.1 Gaussian

As an effective noise minimization technique, the Gaussian filter is often referred to as a "shaped smoothing filter," although the blurring achieved is much less pronounced than the latter. A typical 3×3 Gaussian filter is:

1	2	1
2	4	2
1	2	1

Unlike a smoothing filter, the PUI weighting is greater than 1 and also greater than any of the neighboring pixels. This leads to a stronger inclination to the original PUI value, which in turn promotes a much weaker smoothing effect. As you would expect, a larger kernel grid invites the influence of pixels farther away, increasing the blurring effect:

1	2	4	2	1
2	4	8	4	2
4	8	16	8	4
2	4	8	4	2
1	2	1	2	1

5.8.2.2 Gradient

Often one of the most interesting (and fun to play with!) filter families is the gradient. This filter is particularly useful when attempting to map intensity variations along a specific axis of the image. An example of a $45°$ gradient filter is:

0	-1	-1
1	0	-1
1	1	0

This filter example seemingly "sweeps" through the image from the top right corner, highlighting variations in intensity, in the same manner as a first-order derivative. If you were to imagine this sweeping, as the filter detects an increase in intensity, it is marked lighter and decreases are marked darker (Figure 5.65). Although I have described this filter as having a "direction," all noncumulative filters are purely directionless — the new image is built from artifacts of the source image only and therefore mathematically filtering the pixels in differing directions (or in a random order) has no effect.

Another useful feature of the gradient filter is your ability to increase the apparent thickness of the edges detected. As you increase the size of your kernel, the thickness also increases (Figure 5.66).

As you can see from the 7×7 kernel example, the image has become quite distorted as the intensity gradients are more highlighted from the background, so the value of increasing it to this level is questionable (depending on the software application's purpose). In the previous example, the 5×5 kernel looks to be a good choice and the effect of the filter is even more apparent when the negative of the image is taken (Figure 5.67).

(a)

(b)   (c)

FIGURE 5.65 (a) Original shrub leaves image. (b) 45° Gradient filtered image. (c) 125° Gradient filtered image.

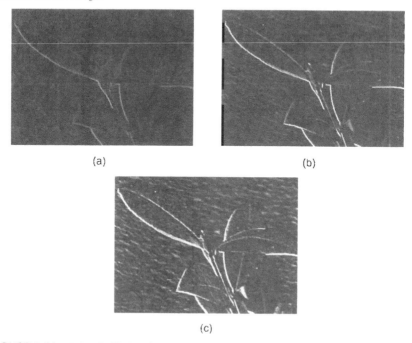

(a)   (b)

(c)

FIGURE 5.66 (a) 3 × 3 45° Gradient filtered image. (b) 5 × 5 45° Gradient filtered image. (c) 7 × 7 45° Gradient filtered image.

FIGURE 5.67 5 × 5 45° Gradient filtered negative image.

5.8.2.3 Laplacian

The easiest way to understand a Laplacian filter is to consider it an omni-directional gradient filter — rather than defining a "direction" of the contour mapping, all contours are enhanced. Consider a 3 × 3 Laplacian filter to be defined as:

A	B	C
D	x	D
C	B	A

If $x = 2(|A| + |B| + |C| + |D|)$, the filter is a very efficient edge detector, increasing the intensity in areas of high contrast and conversely decreasing in areas of low contrast. An example kernel is:

-1	-1	-1
-1	8	-1
-1	-1	-1

Therefore, the filtered image will contain bright lines at well-defined edges, (Figure 5.68). Similarly, increasing the size of the kernel enhances the edge detection lines (Figure 5.69).

On the other hand, if $x > 2(|A| + |B| + |C| + |D|)$, then the filter works in the same way as described above, except that each original PUI is retained (therefore, the original image is retained). This technique can be useful when a point of reference is difficult to find in the filtered image (Figure 5.70):

-1	-1	-1	-1	-1
-1	-1	-1	-1	-1
-1	-1	29	-1	-1
-1	-1	-1	-1	-1
-1	-1	-1	-1	-1

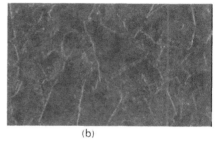

(a)   (b)

FIGURE 5.68 (a) Original crushed terra cotta image. (b) 3x3 Laplacian filtered image.

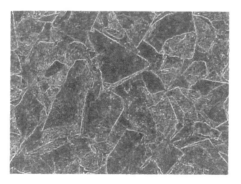

FIGURE 5.69 5 × 5 Laplacian filtered image.

(a)

(b)

FIGURE 5.70 (a) Original crushed terra cotta image. (b) 5 × 5 Laplacian filtered image with PUI.

5.8.2.4 Smoothing

A smoothing filter is simply an averaging filter, blurring out features, details and object shapes and is the most often used to, often inappropriately, minimize noise. The smoothing kernel assuages the variations in image intensity in the pixel neighboring the PUI. A simple smoothing kernel is:

0	1	0
1	0	1
0	1	0

As you can see, no weight is placed on the PUI — its calculated value is simply an average of those around it, which all have the same weight, resulting in a blurred image. As you would expect, a larger active kernel increases the blurring, as the distant pixels begin to affect the resulting value of the PUI (Figure 5.71(a,b)).

5.8.3 CREATING YOUR OWN FILTER

Although the four general-purpose filter families provided with the Vision Toolkit are quite useful, they are by no means the only ones available. If you are unable to find a pre-defined filter to fit your needs, designing a custom one is very easy. You can simply wire your kernel as a two-dimensional SGL array into IMAQ Convolute (Figure 5.72).

You can also use a Kernel string parser called IMAQ BuildKernel to construct a kernel array from a string. The code shown in Figure 5.73 yields the same result as the example in Figure 5.72. This technique can be very useful when reading kernels from a text file, perhaps edited by a user not familiar with LabVIEW, or allowing filters to be altered as required after your application has been built, permitting you to distribute filters in a plug-in style.

It is good to model a filter before coding it, to get it right before opening LabVIEW. Find an excellent Web-based image convolution filter kernel utility at http://www.cs.brown.edu/exploratory/research/applets/appletDescriptions/filterKernel/home.html (Figure 5.74). This Java applet, developed by John F. Hughes and Da Woon Jung at Rhode Island's Brown University, allows you to select an image, tweak standard kernels or define your own and then view the effect your filter will have. Using this utility could save you a lot of development time.

(a)   (b)

FIGURE 5.71 (a) Original King of Hearts image. (b) 5 × 5 smoothing filtered image.

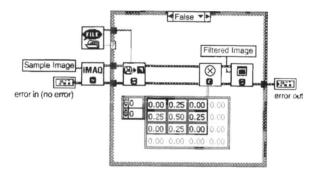

FIGURE 5.72 Constant custom filter convolution kernel — wiring diagram.

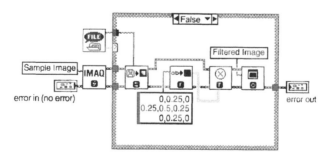

FIGURE 5.73 Constant filter kernel parser — wiring diagram.

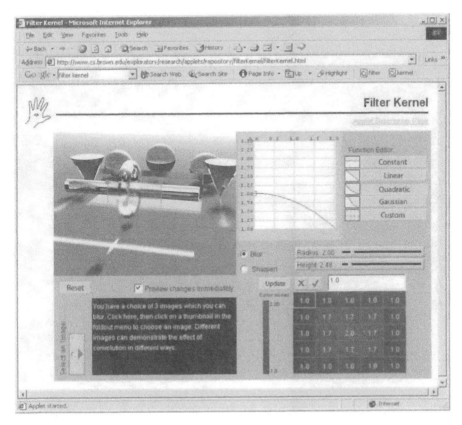

FIGURE 5.74 Online filter kernel applet.

6 Morphology

When applied to vision, morphology (from the Greek word *morph*, meaning shape or form) refers to the alteration of an image using computer routines. A more-common use for the term is the changing of an image in graduated steps (similar to the "morphing" of an image of a baby to an adult), although this is a particular type of morphology.

Occasionally, acquired images are not quite what we are looking for. They may contain particle noise, or objects may have holes in them that we do not want our vision routines noticing. When the acquired image is not suitable in these respects, then you need to morph.

6.1 SIMPLE MORPHOLOGY THEORY

Morphological operations are generally neighborhood based (the new value for the pixel under inspection (PUI) is determined from the values of its neighboring pixels). The two main types of binary morphological operations are *erosion* and *dilation* (other morphological operations exist, but are generally combinations of the two). Morphological operations are very similar to filters, except the kernels used are dependent on the original value of the PUI.

6.1.1 DILATION

Similar to the biological condition when referring to the human iris, dilation refers to the spatial expansion of an object with the potential of increasing its size, filling holes and connecting neighboring objects. Consider an image with the following elements:

NW	N	NE
W	*x*	E
SW	S	SE

Simple dilation occurs when the following procedure is applied: if the PUI is 1, then it retains its value; if it is 0, it becomes the logical OR of its cardinal (North, South, East and West) neighbors:

$$x_{New} = \begin{cases} 1 & x_{Old} = 1, \exists(N,E,S,W) = 1 \\ 0 & otherwise \end{cases}$$

An example is shown below. Note that any PUIs with an original value of 1 are retained, small holes are filled and boundaries are retained:

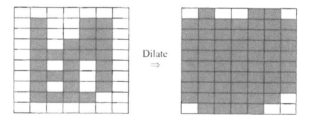

6.1.2 EROSION

Similar to a small island in the middle of a river during water flow, erosion wears away the edges of image features, decreasing its size, and potentially opening up holes in the object. Consider again the previous image with the following elements:

NW	N	NE
W	x	E
SW	S	SE

Simple erosion occurs when the following is applied: if the PUI is 0, then it retains its value; if it is 1 and all of its cardinal neighbors are also 1, then set the new pixel value to 1 (otherwise set it to 0):

$$x_{New} = \begin{cases} 0 & x_{Old} = 0 \\ 1 & x_{Old} = 1 \text{ and } \wedge(N,E,S,W) = 1 \end{cases}$$

An erosion example is shown as follows:

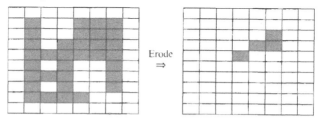

6.1.3 CLOSING

To "close" small holes and other background features in an image, all you need to do is first dilate the source image, and then erode it:

$$I_{Closed} = Erode\{Dilate(I_{Source})\}$$

Although performing two seemingly opposite morphs on the source image may suggest that it will result in the destination image being the same as the source, this is not the case. Original image data is lost on executing the first operation, and hence the original image cannot be recovered.

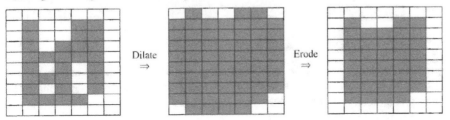

As you can see, the small inner holes were filled, as was a portion of the upper inlet.

6.1.4 OPENING

Opening an image (expanding holes and other background features in an image) requires the opposite sequence to closing: first erode the source image, and then dilate it:

$$I_{Opened} = Dilate\{Erode(I_{Source})\}$$

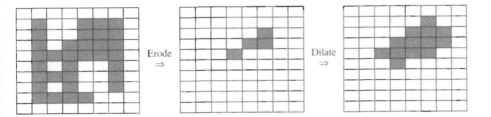

The opened image may not look much like its source, although the thinner portions of the object have been successfully opened, and only the bulk area remains.

6.1.5 NONBINARY MORPHOLOGY

Although the simplest method of morphology is to first threshold an image, creating a binary image (a pixel is either light or dark — there is no in-between), other methods exist such as gray and color morphology.

Gray morphology is achieved by considering each of the neighboring colors as binary couples, and performing binary morphology between them. This process is

FIGURE 6.1 Simple morphology — erosion and dilation.

repeated for $n_{colors} -1$ (i.e., the number of color neighboring couples), and then these morphed images are weighted back using their respective gray levels (each morphed binary is typically promoted to the next highest gray level) and recombined into an image with $n_{colors} -1$ colors.

6.2 PRACTICAL MORPHOLOGY

Simple dilation, erosion, opening and closing can be easily achieved using either IMAQ Morphology or IMAQ GrayMorphology[1] (Figure 6.1).

The morphological function performed on the source image in Figure 6.1 was performed using the code shown in Figure 6.2. Similarly, both opening and closing

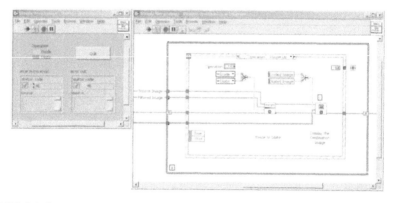

FIGURE 6.2 Simple morphology: erosion and dilation — wiring diagram.

can be achieved using the same VIs. As you can see in Figure 6.3, the opening routine discards the small specular objects around the main object, whereas the closing routine attempts to include the specular data to the east of the circle, effectively closing the spaces between the solid object and close outlying objects (Figure 6.4)

---

[1] These vision routines presume that the object of an image is the brightest section, and the background is the darkest; therefore, you may need to invert an image before attempting practical morphology.

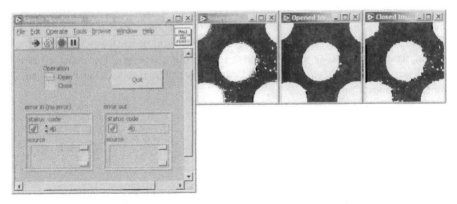

FIGURE 6.3 Simple morphology — opening and closing.

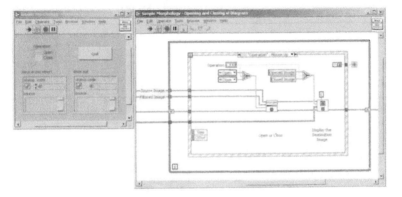

FIGURE 6.4 Simple morphology: opening and closing — wiring diagram.

6.3 THE STRUCTURING ELEMENT

You can change the way the morphing routines transform an image by defining a custom *structuring element*. A structuring element is a binary mask that defines which neighboring pixels are considered when performing a morphological function. The previous examples have been executed using a standard structuring element:

0	1	0
1	1	1
0	1	0

You can define custom structuring elements by passing in a two-dimensional 32-bit array, as long as both the *n* and *m* sizes are equal and odd (Figure 6.5). Custom structuring elements can be very useful if you need to erode or dilate an object from a particular angle (similar to angular filtering). The matrix below shows some examples of nonstandard structuring elements and their effect on a morphing operation.

Source Image	Structuring Element	Dilated Image	Eroded Image

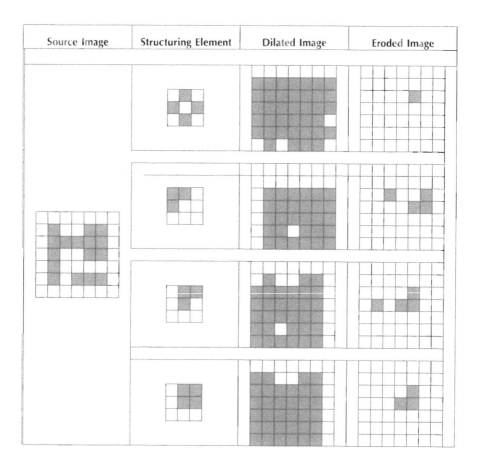

6.4 SPECIFIC MORPHOLOGICAL FUNCTIONS

6.4.1 PARTICLE REMOVAL

Removing unwanted particles is one of the most-common uses of morphology. Perhaps you have an acquired image that contains several objects of varying size, and need to count the number larger than a set point, or maybe the smaller objects

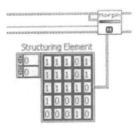

FIGURE 6.5 Custom structuring element.

will be discarded in a future production process, so should be ignored when building statistical datasets. IMAQ Filter Particle is a very powerful VI in the Vision Toolkit, which can be used for particle removal (Figure 6.6).

The Selection Values input is an array of clusters (Figure 6.7). The cluster defines the particle removal method (Table 6.1). The cluster is placed inside an array, and it permits multiple selection criterion to be defined. For example, you can perform a particle removal that removes particles smaller than a specific area, and those that do not confirm to a defined circularity, all within one VI (each particle removal step executes in the same order as they are defined in the array).

A simple particle removal example is shown in Figure 6.8. In this example, the selection values array has only one element, with the cluster values defined as shown in Table 6.2. This was achieved using the code shown in Figure 6.9.

IMAQ BasicParticle was used to determine the number of particles both before and after the routine was completed. This VI is also useful when you are initially determining the lower and upper values of the particle removal, as it also returns an array that describes each of the particles detected (Figure 6.10). The clusters return the pixel and calibrated areas (if calibration data exists) of each

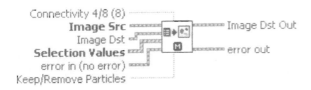

FIGURE 6.6 IMAQ Filter Particle.

detected particle, and its global rectangle (a rectangle that is parallel to the image's border and bounds the particle) (Figure 6.11).

6.4.2 Filling Particle Holes

Once you have removed any unwanted particles from the image, a good next step is to fill any holes in the images objects. Filling holes gives you solid image objects, which are more easily counted and measured (Figure 6.12). IMAQ FillHole accepts binary source images and changes the intensity values of any detected hole pixels to 1

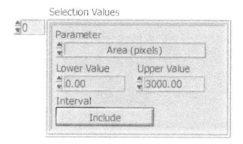

FIGURE 6.7 Morphology selection values cluster.

TABLE 6.1
Selection Values Cluster

Cluster Item	Description
Parameter	Defines the type of removal to be performed. Examples include *area, center of mass position, segment lengths, orientation* and *circularity*.
Lower value	The lower value which the parameter is based on. For example, when using an *area* parameter, the lower value is the smallest area on which to base the area removal routine: if using the include interval, particles smaller than the lower value are discarded.
Upper value	The upper value which the parameter is based on. For example, when using an area parameter, the upper value is the largest area on which to base the area removal routine: if using the include interval, particles larger than the lower value are discarded.
Interval	Whether to include the range between the upper and lower values, or exclude them when applying the *parameter* routine.

(maximum intensity in a binary image). As seen in Figure 6.12, holes ttouching the edge of the image are not filled, as it is impossible from the information provided to determine whether they are holes or outside edges of irregularly shaped objects (Figure 6.13).

6.4.3 Distance Contouring

6.4.3.1 IMAQ Distance

Distance contouring is a neat function that visually maps the relative distance of a pixel to its border's edge. More strictly, distance contouring encodes a pixel value of a particle as a function of the location of that pixel in relation to the distance to the border of the particle. As you might expect, the source image must have been created with a border size of at least 1, and be an 8-bit image. Using IMAQ Distance, you can also define whether the routine uses a square or hexagonal pixel frame during the transformation (Figure 6.14).

    IMAQ Distance is a very simple VI to use, as shown in Figure 6.15.

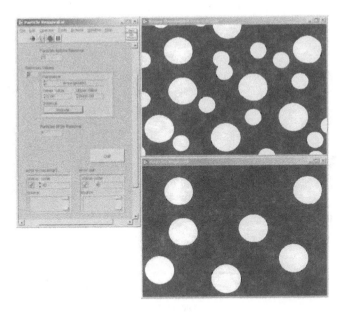

FIGURE 6.8 Particle removal.

TABLE 6.2
Selection Values Cluster Example

Cluster Item	Value	Description
Parameter	Area (pixels)	Particle removal is based on the pixel area of detected particles (a calibrated area routine is also available)
Lower value	0.00	
Upper value	3000.00	
Interval	Include	Process the routine using the values between and including the lower (zero pixels area) and upper (3000 pixels area) values

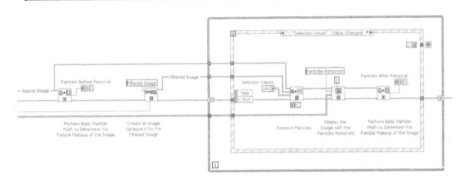

FIGURE 6.9 Particle removal — wiring diagram.

Basic Reports

Area (pixels)			Area (pixels)			Area (pixels)		
9030			18318			18266		
Area (calibrated)			Area (calibrated)			Area (calibrated)		
9030.00			18318.00			18266.00		
Global Rectangle☐			Global Rectangle☐			Global Rectangle☐		
380		x1Left	808		x1Left	118		x1Left
0		y1Top	62		y1Top	64		y1Top
533		x2Right	961		x2Right	273		x2Right
73		y2Bottom	213		y2Bottom	213		y2Bottom

FIGURE 6.10 IMAQ basic particle basic reports.

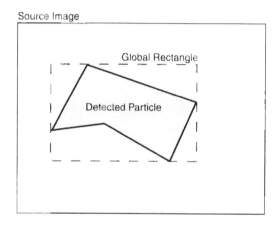

FIGURE 6.11 The global rectangle.

6.4.3.2 IMAQ Danielsson

Although IMAQ Distance is useful for fast distance contouring, as seen from Figure 6.14, the transform is far from accurate. A much more accurate, yet slower approach is to use IMAQ Danielsson (Figure 6.16). The *Danielsson distance map* algorithm, which was devised by Erik Danielsson, is based on the Euclidian distance map. The map values are coordinates giving the *x* and *y* distances to the nearest point on the boundary, and are stored as complex valued pixels. Then the absolute function is used to transform these into the radial distance. More information about the Danielsson distance map can be found in the paper "Distance Maps for Boundary Detection and Particle Rebuilding," by Frédérique and Lefebvre (http://www.ici.ro/ici/revista/sic97_3/GUY.html). IMAQ Danielsson is even easier to use than IMAQ Distance, as it is not necessary to specify whether to use square or hexagonal pixel frames (Figure 6.17).

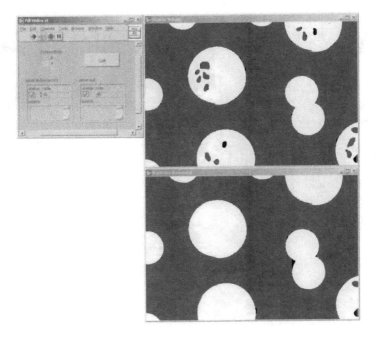

FIGURE 6.12 Fill holes.

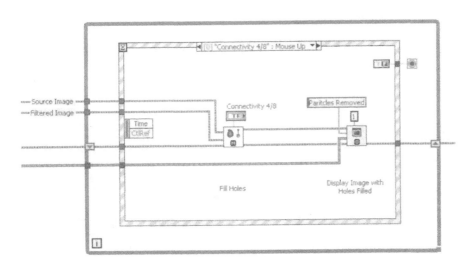

FIGURE 6.13 Fill holes — wiring diagram.

6.4.4 BORDER REJECTION

Objects are often intersected by the image border, and are therefore difficult to characterize. Figure 6.18 includes an irregular object that could be confused with a circle, and two circles with faults that could be assumed as holes. Although assump-

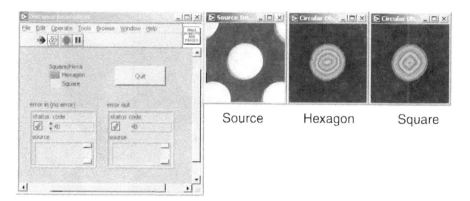

Source   Hexagon   Square

FIGURE 6.14 Distance example.

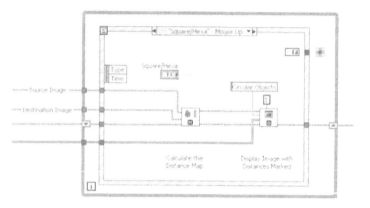

FIGURE 6.15 Distance example — wiring diagram.

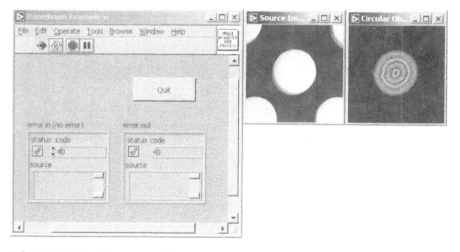

FIGURE 6.16 Danielsson example.

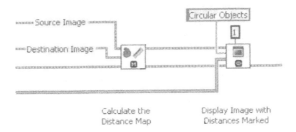

FIGURE 6.17 Danielsson example — wiring diagram.

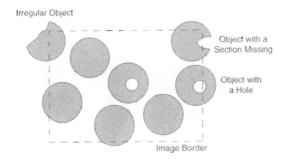

FIGURE 6.18 Irregular objects touching the image border.

tions could be made to extend these fragments to full objects, not enough information is present to accurately do so. This is why it is often appropriate to ignore any objects that are touching the image border, and perhaps consider them in future acquired frames if the objects are moving.

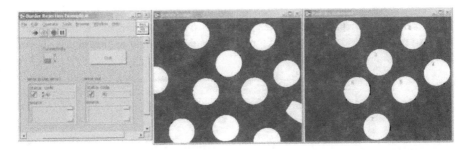

FIGURE 6.19 Border rejection example.

The example shown in Figure 6.19 counts and labels the total number of objects that are wholly within the image. This was achieved using IMAQ RejectBorder, which has very few inputs: the source and destination image data spaces and the connectivity (Figure 6.20).

6.4.5 FINDING AND CLASSIFYING CIRCULAR OBJECTS

If the image contains circular objects that you want to classify, a very simple VI to use is IMAQ Find Circles. This VI will search for circles in the image and return

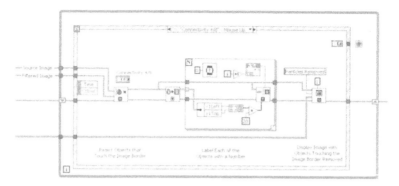

FIGURE 6.20 Find and classify circular objects — wiring diagram.

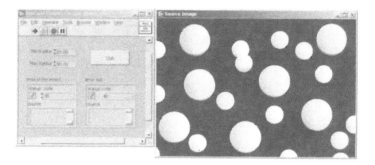

FIGURE 6.21 Find and classify circular objects .

all sorts of information about them, including the number of circles detected, and their respective center positions, radii and calculated area (Figure 6.21).

IMAQ Find Circles uses a *Danielsson distance map* to determine the radius of each object and can also separate perceived overlapping circles. This is implemented using the code shown in Figure 6.22, and an example application is shown in Figure 6.23.

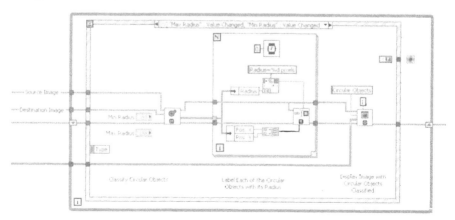

FIGURE 6.22 Find and classify circular objects — wiring diagram.

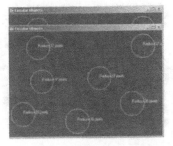

Minimum = 30, Maximum = 40

Detected circles with radii between 30 and 40 pixels are included in the search; therefore, the smaller objects are disregarded.

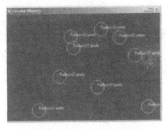

Minimum = 10, Maximum = 30

Detected circles with radii between 10 and 30 pixels are included in the search; therefore, the larger objects are disregarded.

Minimum = 10, Maximum = 40

All of the circular objects are detected, as the search radii range is between 10 and 40 pixels. Note that the two small circles that were overlapping have been separated and characterized correctly.

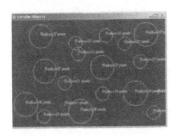

Minimum = 1, Maximum = 40

Decreasing the lower limit of the radii search range to 1 pixel has detected two extra circles, both of which are actually noise in the image.

FIGURE 6.23

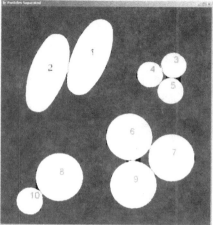

FIGURE 6.24 Particle separation example.

6.4.6 PARTICLE SEPARATION

If two or more particles are touching or overlapping in an image, you can separate them using IMAQ Separation. This VI performs a series of morphological erosions and then reconstructs the image based on any created object divisions (Figure 6.24).

If an isthmus is broken or removed during the erosion process, the particles are reconstructed without the isthmus. The reconstructed particles generally have the same size as their respective initial particles, although this can fail when the number of erosions is high. IMAQ Separation is simple to use (note that the majority of the code in Figure 6.25 is to label each particle with a number).

If the number of erosions is small, IMAQ BasicParticle can be used to determine the area of the newly separated particles; if the number of erosions is high, particle counts andglobal rectangle measurements are generally still quite accurate.

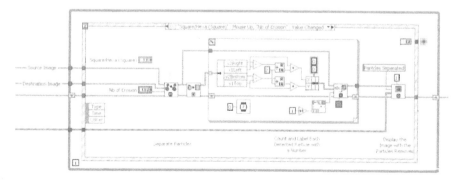

FIGURE 6.25 Particle separation example — wiring diagram.

6.5 CASE STUDY: FIND AND CLASSIFY IRREGULAR OBJECTS

A user recently posted a question on the National Instruments Developers' Zone discussion forum:

> I want to read a barcode that is [a small part of] a big image.... I am not able to read the barcode as I have to give the correct ROI, which in my case has to be set dynamically.

The image in question was indeed large (1600 × 1400 pixels), and the position that the barcode label was attached to the object was undefined. The approach used was divided into several steps, as shown in Table 6.3.

TABLE 6.3

VI	Function
IMAQ Resample	Resize the image to 800 × 600 pixels (the extra image information was not required)
IMAQ Threshold	Threshold the image to highlight the label (it is one of the brightest areas of the source image)
IMAQ FillHole	Fill any holes in the objects that remain (including the black numbers and bars of the actual barcode)
IMAQ Particle Filter	Filter out any objects that have calibrated areas outside of a defined range (this removes objects that do not have the same area as the barcode label)
IMAQ BasicParticle	Find the Bounding Rectangle of the remaining object (the barcode label)
IMAQ Extract	Extract the barcode label from the source image
IMAQ Read Barcode	Parse the label for its value

The VI was coded as shown in Figure 6.26. As IMAQ BasicParticle was used to find the bounding rectangle of the barcode label, the position of the label (as long as it was completely within the source image) was irrelevant (Figure 6.27).

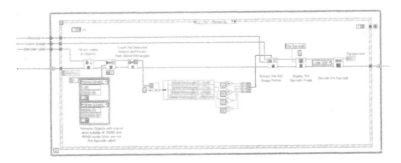

FIGURE 6.26 Search for label and read barcode — wiring diagram.

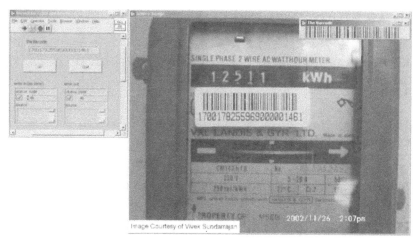

FIGURE 6.27 Search for label and read barcode.

6.6 USER SOLUTION: DIGITAL IMAGING AND COMMUNICATIONS IN MEDICINE (DICOM) FOR LABVIEW

Chris Reed
Development Engineer
Department of Medical Physics
Royal Perth Hospital, Australia
Tel +61-8-9224-2953/Fax +61-8-9224-1138
e-mail: Christopher.Reed@health.wa.gov.au

The DICOM Encoder is at the center of a suite of software tools designed for the acquisition and analysis of ultrasound images in cardiovascular research, although it can be easily adapted to any of the standard medical areas such as angiography and nuclear medicine. The main function of the DICOM Encoder is to convert an analog video signal (PAL or NTSC) into a DICOM-compatible digital file and log it to disk. This is achieved in real-time and at the full frame rate of the incoming video signal, using the National Instruments range of image acquisition cards. The highest quality signal is captured directly from the ultrasound machine, although previous studies on analog tapes can also be converted into digital DICOM format through a standard VCR. If required, a PCI expansion box can be used with a laptop to provide a portable solution for video capture and archiving. A standard DAQ card can also be used to acquire physiological data (e.g., pressure), but the latest approach is to display this information in the actual image, providing a cheaper solution without being concerned with synchronizing the acquired data with the video sequence.

The DICOM Encoder was developed to bridge the gap between the old analog technology and the new digital world of DICOM and overcomes the following problems:

- *Hardware limitations*: ultrasound machines typically have limited ram and hard drive capacity, unlike that of a high-performance pc, greatly extending the available capture time of individual studies from a minute or two to over half an hour or more. The accessing and automated archiving of such large files is also much quicker and easier.
- *Pulse wave (pw) mode*: common ultrasound machines are unable to capture and store continuous video in pw mode. As the dicom software captures the raw signal directly from the ultrasound probe, there are no limitations on which mode the ultrasound unit is in. This feature is essential for techniques including blood flow analysis, which requires the doppler velocity trace.
- *Consistent frame rates*: the dicom system can guarantee 25 fps (pal), 30 fps (ntsc) sustained capture rates, regardless of image compression or size.
- *Image resolution/quality*: using the "camera" quality analog output from the ultrasound machines, it is possible to greatly improve the quality of

the digital image stored to file as opposed to the traditional analog tape, currently considered the "gold standard."

- *File formats*: the dicom file format offers random access to the data, a worldwide standard file format and no image quality degradation over time.

6.6.1 THE HARDWARE

Figure 6.28 shows the hardware system overview.

6.6.2 THE SOFTWARE

DICOM was developed using a number of software tools:

- LabVIEW 6.1
- IMAQ Vision Toolkit 6.0
- Intel JPEG Library v1.51
- DicomObjects 4.1
- LabVIEW Image Library v1.01

After a system frame rate hardware/software test is completed, the user is then allowed to acquire a study. Relevant study information (patient name, hardware, medical details, etc) have already been entered and encoded in the DICOM Header Interface (Figure 6.29).

The Settings interface allows the user to select the image format (color/monochrome) and has other default settings that are used by other modules in the program (Figure 6.30).

Initially, the user can perform an auto exposure on the image to set the optimal brightness level, or can manually adjust the white/black reference levels. When the *Record* button is pressed, acquired images are encoded into JPEG format, displayed and logged to disk simultaneously. The study details (file size, number of frames, frame size and study length) are also updated. When the study is complete, the user can calibrate the image to convert screen pixels into a known length, velocity or any other physiological signal (pressure, ECG, etc.). All this information is embedded into the file according to the format of the DICOM Standard.

Although typical commercial medical devices may occasionally store a file as large as 20 MB, they are generally only a couple of megabytes. However, research studies can typically last up to 20 minutes, thus creating a single data file that could be in excess of 1.2 GB. Standard DICOM files of this size can take several minutes to save after acquisition, so a specialized *fast save* format was developed, allowing a study of any size to be ready for review immediately after the acquisition is completed. This new format is not strictly DICOM compatible, although a conversion utility is provided to convert the fast save data to the standard format.

Any DICOM files can be read and displayed in the *Edit/Review* Interface, and all the study information is displayed in a separate window when the *Details* button is pressed. A horizontal pointer slide is used to display any frame in the

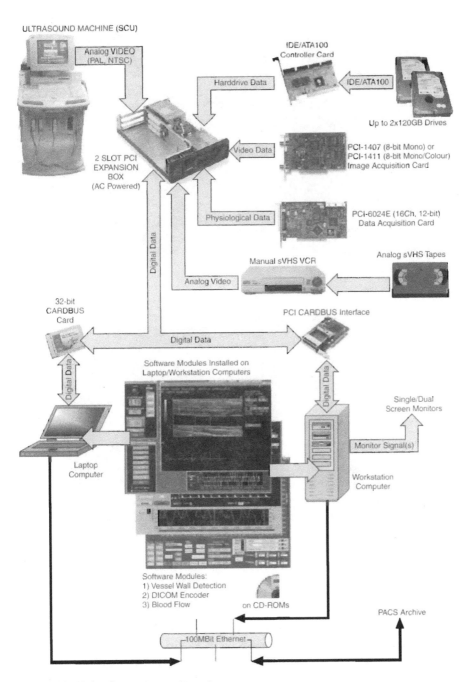

FIGURE 6.28 Hardware System Overview.

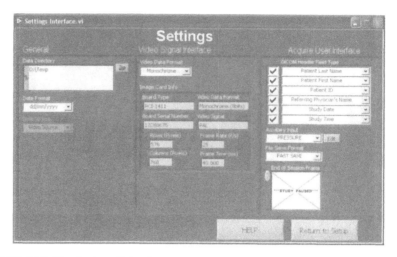

FIGURE 6.29 The *Settings* dialog box.

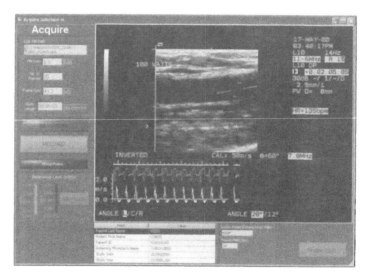

FIGURE 6.30 The *Acquire* user interface.

study, and an XY Graph displays the length of the study and markers for the time the user pressed paused during acquisition. It is also used to mark the start and end of a sequence, which can be converted into an MPEG or AVI movie using any of the installed system codecs. The user also has a range of VCR-style buttons to review the study, with the *Play Rate* control allowing rates up to eight times normal speed. By changing the *Search Mode* from *Normal* to *Index*, the user can instantly jump to the start of each session using the pause markers as a reference (Figure 6.31).

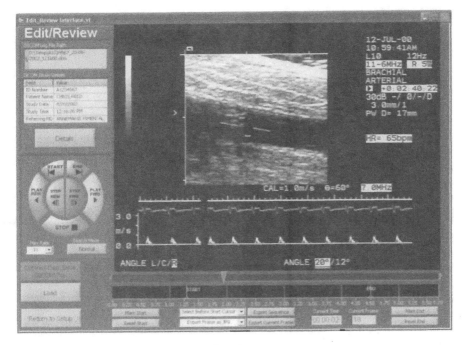

FIGURE 6.31 The *Edit* and *Review* user interface.

6.6.3 SUMMARY

The major lesson learned from developing this package was to use established software wherever possible to reduce development time. Using LabVIEW and IMAQ simplified development and provided a range of advanced features that linked the hardware and software together, albeit at a high cost. Using these two tools made this part of the software relatively easy to accomplish. The most difficult aspect of the development was the understanding and interfacing to the DICOM Standard, which was achieved using an ActiveX toolkit and has so far proved remarkably robust in the latest version of LabVIEW, considering the problems in earlier versions.

6.6.4 ACKNOWLEDGMENTS

- Dr. David Harvey from Medical Connections, for his DicomObjects ActiveX component, which handles all aspects of the DICOM Standard and his expert knowledge of the field (see www.medicalconnections.co.uk for more information).
- Richard Weatherill for his help in developing the Encoder software.
- Holger Lehnich from the Martin-Luther-University Halle-Wittenberg Medical Faculty for the use of his Motion JPEG Encoder and its further development for the DICOM application.
- Icon Technologies for their LabVIEW Image Library and valued support as a National Instruments Select Integrator/Alliance Member. See their website at http://www.icon-tech.com.au for more information.

7 Image Analysis

Image analysis is a nondestructive method of computing measurements and statistics based on the intensity values of an image's pixels, and their respective spatial location within that image. Analyzing an image can give you insight to objects that it might contain, and allows you to perform measurements on those objects, or verification on their presence.

7.1 SEARCHING AND IDENTIFYING (PATTERN MATCHING)

If you are looking for a particular object within an image, IMAQ Count Objects is not really going to help you, unless you specifically know that the only objects in your images are the type that you are looking for. What you will need is a template of what you are looking for, a list of degrees of freedom that you are willing to accept, and a routine to search for the template within the acquired image.

Consider the image of a PCB shown in Figure 7.1. This PCB design has three similar circuits laid out on the board in parallel. If you were looking for the existence of a particular component, an image of that component (or a suitable portion of it) would form your search template (Figure 7.2). By performing a grayscale pattern match, you are able to identify possible instances of the template (Figure 7.3).

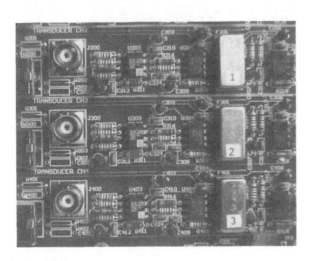

FIGURE 7.1 Sample PCB.

7.1.1 The Template

First you need to define what you want the search routine to look for — the template. It is important to choose a template with high contrast, as the Vision Toolkit search routine does not simply try to find exact matches to your template's grayscale intensity matrix. Instances of the template in the source image under different ambient lighting conditions and with different background information can often be detected as well (Figure 7.4).

Another guideline is to try to include only the information you need in a template, as including image data around the required object template can cause the search routines to score a match poorly based on what may be irrelevant background data (Figure 7.5).

You can define a template in one of two ways: the user can define a template by drawing an ROI (region of interest) over an existing image, or a template can previously exist as an image file. When you save a template as a PNG file, several other details are embedded in the header of the file to aid in future pattern matching. These templates will open as standard PNG files when accessed using imaging software (e.g., PaintShop Pro), but if you subsequently save an image to the PNG

FIGURE 7.2 Sample template

FIGURE 7.3 Sample template search — 3 matches found.

format, the extra information will be stripped. The image file will be valid, but the template portion of it will be lost.

7.1.2 The Search

There are several methods of performing a pattern match, and below is some limited theory on the main methods used in the vision industry.

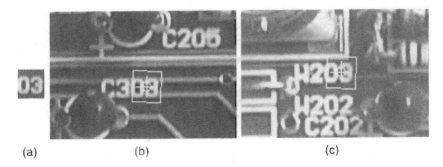

FIGURE 7.4 (a) High contract template. (b) Background variance. (c) Lighting variance.

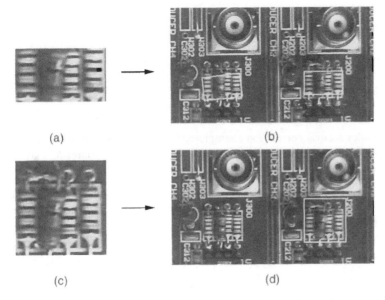

FIGURE 7.5 (a) Minimal template (b) Two instances found. (c) With background data. (d) One instance found.

7.1.2.1 Cross Correlation

The mathematical process of image cross correlation is a simple one. Conceptually, the template is laid over the source image, and the intensity values for each corresponding pixel are individually multiplied, and then all of them are summed to produce a single correlation value. The template is then moved one pixel, and the process is repeated until the whole source image has been covered, and a matrix of correlation values has been created. The correlation value matrix is then scanned for its highest value; this position generally refers to the position in the source image that most closely matches the template. Of course, the correlation matrix can have several high values in it, corresponding to several instances of the template in the source image.

To explain the concept of cross correlation further, consider a source image matrix of size a × b, and a template image matrix of size c × d (when c ≤ a and d ≤ b). If both the source image and template are normalized, the cross correlation matrix is populated using the following equation:

$$CrossCorrelationMatrix_{i,j} = \sum_{x=0}^{b-1} \sum_{y=0}^{a-1} \left(Template_{x,y} \right) \left(Source_{(i+x),(j+y)} \right)$$

If the images have not been normalized, then we must normalize them by dividing each element in the respective images by the square root of the sum of its squares:

$$CrossCorrelationMatrix_{i,j} = \frac{\sum_{x=0}^{b-1} \sum_{y=0}^{a-1} \left(Template_{x,y} \right) \left(Source_{(i+x),(j+y)} \right)}{\sqrt{\left(\sum_{x=0}^{b-1} \sum_{y=0}^{a-1} \left(Template_{x,y} \right)^2 \right) \left(\sum_{x=0}^{b-1} \sum_{y=0}^{a-1} \left(Source_{x,y} \right)^2 \right)}}$$

Consider a cross correlation example with a 3 × 3 template matrix (Table 7.1). To make the example simpler, let us assume that both the source S_{xy} and template T_{xy} images have been normalized. Performing the cross correlation between the images yields:

TABLE 7.1

S_{00}	S_{01}	...	...	S_{0b}
S_{10}	...	...	...	S_{1b}
...	...	...	...	...
...	...	...	...	...
S_{a0}	S_{a1}	...	...	S_{ab}

T_{00}	T_{01}	T_{02}
T_{10}	T_{11}	T_{12}
T_{20}	T_{21}	T_{22}

Some values used in the cross correlation are undefined (i.e., those cells on the right and bottom of the matrix), and are consequently set to zero for calculations. This makes finding matches of a partial template near the edges unlikely.

As you might imagine from the equation matrix shown previously, performing a cross correlation on large images can take some time, although increasing the size of the template will make the routine even slower. Also, if the rotation of the template is not known, the cross correlation will need to be repeated at a range of angles to achieve rotation-invariant matches.

7.1.2.2 Scale Invariant and Rotation Invariant Matching

One of cross correlation's biggest flaws is its inability to match objects in a source image that are either a different size to the template, or have been rotated. If this

$$\begin{array}{l}(T_{00} \times S_{00}) + (T_{10} \times S_{10}) + (T_{20} \times S_{20}) + \\ (T_{01} \times S_{01}) + (T_{11} \times S_{11}) + (T_{21} \times S_{21}) + \\ (T_{02} \times S_{02}) + (T_{12} \times S_{12}) + (T_{22} \times S_{22})\end{array} \qquad \begin{array}{l}(T_{00} \times S_{10}) + (T_{10} \times S_{20}) + (T_{20} \times S_{30}) + \\ (T_{01} \times S_{11}) + (T_{11} \times S_{21}) + (T_{21} \times S_{31}) + \\ (T_{02} \times S_{12}) + (T_{12} \times S_{22}) + (T_{22} \times S_{32})\end{array} \quad \cdots \quad \begin{array}{l}(T_{00} \times S_{a0}) + (T_{10} \times S_{a+1,0}) + (T_{20} \times S_{a+2,0}) + \\ (T_{01} \times S_{a1}) + (T_{11} \times S_{a+1,1}) + (T_{21} \times S_{a+2,1}) + \\ (T_{02} \times S_{a2}) + (T_{12} \times S_{a+1,2}) + (T_{22} \times S_{a+2,2})\end{array}$$

$$\begin{array}{l}(T_{00} \times S_{01}) + (T_{10} \times S_{11}) + (T_{20} \times S_{21}) + \\ (T_{01} \times S_{02}) + (T_{11} \times S_{12}) + (T_{21} \times S_{22}) + \\ (T_{02} \times S_{03}) + (T_{12} \times S_{13}) + (T_{22} \times S_{23})\end{array} \qquad \cdots \qquad \cdots \qquad \begin{array}{l}(T_{00} \times S_{a1}) + (T_{10} \times S_{a+1,1}) + (T_{20} \times S_{a+2,1}) + \\ (T_{01} \times S_{a2}) + (T_{11} \times S_{a+1,2}) + (T_{21} \times S_{a+2,2}) + \\ (T_{02} \times S_{a3}) + (T_{12} \times S_{a+1,3}) + (T_{22} \times S_{a+2,3})\end{array}$$

$$\cdots \qquad\qquad \cdots \qquad\qquad \cdots \qquad\qquad \cdots$$

$$\begin{array}{l}(T_{00} \times S_{0b}) + (T_{10} \times S_{1b}) + (T_{20} \times S_{2b}) + \\ (T_{01} \times S_{0,b+1}) + (T_{11} \times S_{1,b+1}) + (T_{21} \times S_{2,b+1}) + \\ (T_{02} \times S_{0,b+2}) + (T_{12} \times S_{1,b+2}) + (T_{22} \times S_{2,b+2})\end{array} \qquad \cdots \qquad \cdots \qquad \begin{array}{l}(T_{00} \times S_{ab}) + (T_{10} \times S_{a+1,b}) + (T_{20} \times S_{a+2,b}) + \\ (T_{01} \times S_{a,b+1}) + (T_{11} \times S_{a+1,b+1}) + (T_{21} \times S_{a+2,b+1}) + \\ (T_{02} \times S_{a,b+2}) + (T_{12} \times S_{a+1,b+2}) + (T_{22} \times S_{a+2,b+2})\end{array}$$

issue is to be overcome, the template needs to be rescanned over the source image using different rotations and sizes (variances in both x and y). This can be extremely time consuming; consider performing a cross correlation 360 times just to perform a rotation-invariant match — and that is without subdegree accuracy.

Several techniques exist that can assist in speeding up scale and rotation invariant matching. The most important method is to scan only for variations that are likely to exist. If your part is always the same size, and no spatial distortion exists, then do not scan for size variations. Similarly for rotation variance, if your part will be repeatedly placed at the same orientation, then you do not need to rescan the source image using a range of different angles (cross correlation can typically detect object rotations of around ±5° without rescanning). If you predict that spatial distortions may exist, only scan for those that are likely, as removing scans that are unlikely to occur will decrease the processing time required to complete a search. When objects may be rotated, it is common to place a small coordinate marker on the object that can be easily found and extract the rotation angle from it. Then with the template rotated to the new angle, perform a standard cross correlation.

7.1.2.3 Pyramidal Matching

Although the accuracy of cross correlation is quite good, it is a very slow technique to use. Pyramidal matching uses a similar searching technique, but subsamples both the source image and template to smaller spatial resolutions, effectively reducing the amount of data to be searched by up to 75%. As the resolutions are lower, more care must be taken in defining what is a potential match, as the search is performed

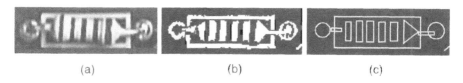

FIGURE 7.6 (a) Original template. (b) Processed template. (c) Geometrically modelled.

on much less data, therefore only areas with very high matching results will be considered as correlating.

7.1.2.4 Complex Pattern Matching Techniques

Instead of attempting to simply match the template's intensity matrix with the intensity matrix of a source image, other techniques are used such as *geometric modeling*, which is the process of breaking down both the source and template images into simply defined geometric objects (Figure 7.6).

The source image is then searched using simpler and much faster binary shape matching routines. Another technique used is *nonuniform* spatial sampling. Consider the following binary image sampled uniformly:

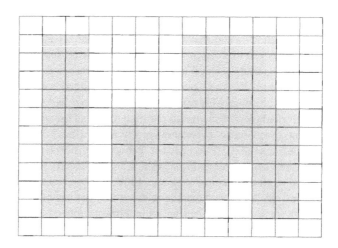

As there are sections of pixels with the same intensities as their neighbors, it can be resampled nonuniformly:

The image is the same, but it is sampled across fewer irregularly shaped pixels.

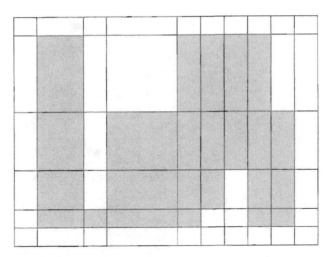

7.1.3 A PRACTICAL EXAMPLE

The example in Figure 7.7 uses `Interactive Pattern Matching Example.vi` (which is on the accompanying CD-ROM) to:

- Prompt the user to load an image to be searched.
- Interactively define a rectangular ROI (the search template).
- Set search parameters.
- Perform a pattern matching.

Once the source image has been loaded, the user defines a ROI, and then presses the *Learn Template* button, which calls the code shown in Figure 7.8.

As you can see, the ROI's global rectangle is returned, and its corresponding underlying portion is extracted from the source image — this forms the search template. The next step is to train the system to recognize it, which is achieved by first defining the learn pattern parameters using `IMAQ Setup Learn Pattern`. This VI simply creates a learn pattern parameter string based on one input: *learn mode*. Learn mode specifies the invariance mode to use when learning the pattern. You can choose from values shown in Table 7.2. Once the template is learned, the next step is to search for it. First, define searching criteria with `IMAQ Setup Match Pattern`. Like `IMAQ Setup Learn Pattern`, this VI simply constructs a match pattern parameter string based on its inputs (Table 7.3). The actual searching process is executed in `IMAQ Match Pattern`. Feed the source image, template and match pattern parameters (from `IMAQ Setup Match Pattern`) into this VI, and it will search the source for the template.

Two further methods of increasing the speed of the search include defining a rectangle within the source image to be searched (if you know that the object is going to be in a particular area of the source image, then there is no sense in searching the whole image), and defining the *number of matches requested*. If you are sure that the image will only have a maximum number of instances of the search object, once they are found it wastes time looking for further instances (Figure 7.9).

The same pattern match could be even faster. Simply rotate the source image and template 180° before performing the search, which causes even less of the source

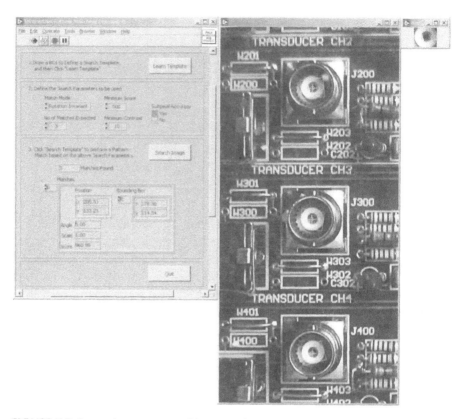

FIGURE 7.7 Interactive pattern matching example.

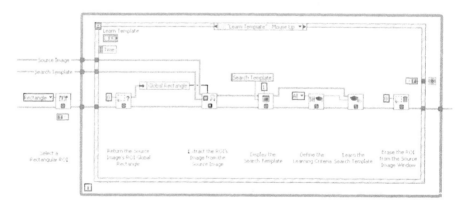

FIGURE 7.8 Learn template — wiring diagram.

image to be searched (the searching routine works from left to right, and then from top to bottom), and the last of the three objects will be closer to the top of the image.

Simple pattern matching is achieved using the code shown in Figure 7.10.

TABLE 7.2
"Learn Mode" Cluster Elements

Value	Description
Shift (default)	Extracts information to specify shift (position) invariant pattern matching
Rotation	Extracts information to specify rotation invariant pattern matching
All	Extracts information to specify both shift and rotation invariant pattern matching

TABLE 7.3
"Setup Learn Pattern" Cluster Elements

Input	Description
Minimum contrast	This defines the minimum contrast you expect to find in the image, which assists the searching when the source image has lighting variances relative to the template image.
Match mode	Sets the invariance mode to either shift (default) or rotation invariant.
Subpixel accuracy	Set whether you want the search routines to use interpolation to attempt searches on a subpixel level. Enabling this option will increase the accuracy of the searching, but will increase the time required to complete the process, therefore should be used only when required.
Rotation angle ranges	If you know that the part will be placed in front of the lens with a rotation of a particular range of angles (e.g., a PCB that will be orientated on a table at either 0, 90, 180 or 270°), they can be defined to effectively eliminate searching for angles that do not exist. Defining appropriate rotation angle ranges can dramatically decrease the search time.

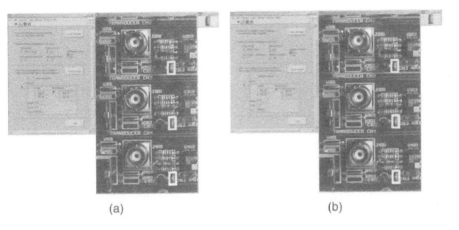

(a)   (b)

FIGURE 7.9 (a) 10 matches expected, 3 found. (b) Three matches expected, three found — faster.

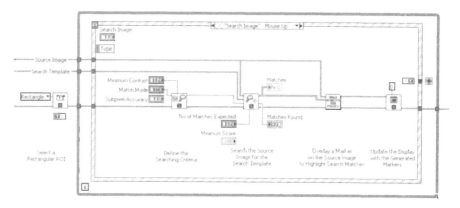

FIGURE 7.10 Search image — wiring diagram.

7.2 USER SOLUTION: CONNECTOR PIN INSPECTION USING VISION

N.D. Buck Smith
Cal-Bay Systems Inc.
3070 Kerner Boulevard, Suite B
San Rafael, CA 94901
Tel (415) 258-9400/Fax (415) 258-9288
e-mail: bsmith@calbay.com
Web site: www.calbay.com

A generic inspection system (Figure 7.11) for verifying defects in pin alignment of standard male connectors was implemented using National Instruments LabVIEW and the Vision Toolkit. The optical system includes a Pulnix TM-200 camera with a Navitar TV-zoom lens and a Fostec ring light. The connector chassis is mounted on a X-Y stage with stepper motors controlled by a PCI-7344 motion controller card and a NI MID-7602 motor drive. Images are acquired using a NI PCI-1408 IMAQ card, and the whole system is shrouded in a fiberglass enclosure mounted on a aluminum frame. The stage has pushpins on X and Y directions to register the connector chassis to the same repeatable position on each load with respect to the camera (Figure 7.12).

The x and y steppers are connected to the NI MID drive, which is connected to the PCI-7344 motion controller in the PCI slot in the PC. Video output of the Pulnix camera is fed to the PCI-1408 in the PC via a BNC cable. The Fostec ring light is mounted on the zoom lens such that it provides uniform illumination for the whole chassis.

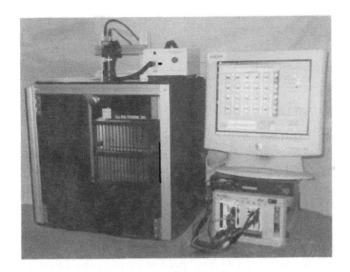

FIGURE 7.11 Connector inspection system demonstration.

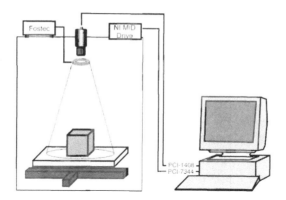

FIGURE 7.12 Connector inspection system diagram.

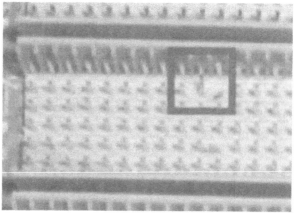

FIGURE 7.13 A bent connector pin.

The connector under test (Figure 7.13) is a 16×7 pin connector, with the bent pin indicated.

7.2.1 THE SOFTWARE

There are two submodes for auto inspection, scan mode and single stepping mode. In Auto Scan mode, the software will automatically go to each connector, unto the number of connectors specified and inspect them for any bent pins or missing pins. The results are shown simultaneously on the display (Figure 7.14). When in Single Stepping mode, the operator can step through each connector manually, with the results displayed after each step.

In Manual Mode, the system can be moved to any connector using the arrow keys. One can also move a specified distance (in inches) by entering the x and y distances, and then pressing GO. The speed of travel can be adjusted from slow to fast. After manually moving to a desired location for inspection, the ROI can be changed by pressing the Set ROI button. Pressing the Process button analyzes the

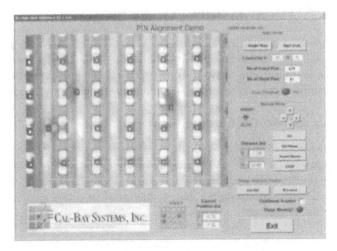

FIGURE 7.14 Front panel of the PIN alignment inspection software.

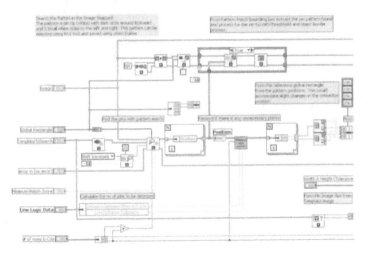

FIGURE 7.15 A portion of the image processing wiring diagram.

image for defective pins in the field of view and displays the results on the front panel in the Results box. The number of bad and good pins is also updated with every image processed.

Shift-invariant pattern matching is used to find the pins and the surrounding areas with two bright spots. Using thresholding and morphology, the bright spots are removed and the coordinates of the pin center are measured and checked for alignment within the limits of the inspection criteria. The overall result of the algorithm is a robust detection routine that eliminates false positives in most lighting conditions (Figure 7.15).

As can be seen from the wiring diagram, an ideal pattern search is implemented using the IMAQ Match Pattern function from the Vision Toolkit.

7.3 MATHEMATICAL MANIPULATION OF IMAGES

If you are not able to find a VI within the Vision Toolkit that will perform the function that you need, and defining a custom filter (see Chapter 5) is not appropriate, then breaking an image into a two-dimensional array of intensity values and performing a custom mathematical process on them may be the only option available.

7.3.1 IMAGE TO TWO-DIMENSIONAL ARRAY

Converting an image to an array is quite simple using IMAQ ImageToArray (Figure 7.16), which parses the image, and returns a two-dimensional array of 8-bit integers that represent the intensity levels across the image (Figure 7.17). IMAQ ImageToArray also has an optional rectangle input, so the user can define a rectangular ROI of intensities to be returned.

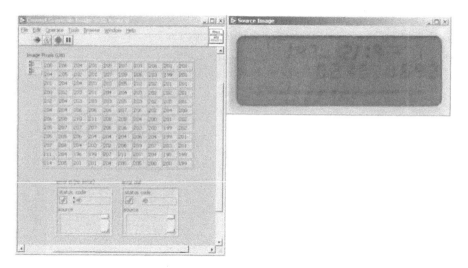

FIGURE 7.16 Convert grayscale image to two-dimensional array.

7.3.2 IMAGE PORTIONS

Rather than returning a two-dimensional array of a complete image or a rectangular ROI, IMAQ GetPixelValue, IMAQ GetRowCol and IMAQ GetPixelLine are useful when you need to extract particular predefined portions of an image. In the example shown in Figure 7.18, the user has selected a pixel in the middle of the image, and IMAQ GetPixelValue has returned the intensity value of that pixel. IMAQ GetRowCol has also been used to return intensity contours of the horizontal row and vertical column that intersect at the PUI, and IMAQ GetPixelLine returns the intensity values along an arbitrary line from the top left corner of the image to the bottom right (Figure 7.19).

IMAQ SetPixelValue, IMAQ SetRowCol and IMAQ SetPixelLine accept new intensity value inputs, allowing you to set the intensity values of image portions (Figure 7.20). As you can see in Figure 7.21, the user can set the new intensity

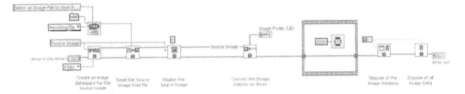

FIGURE 7.17 Convert grayscale image to two-dimensional array — wiring diagram.

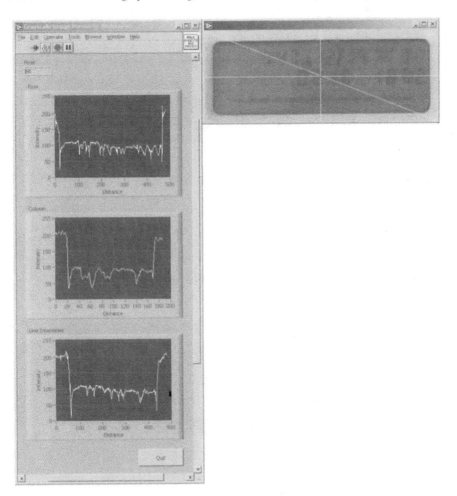

FIGURE 7.18 Grayscale image portions to mathematical values.

value for the PUI. The user can also define a multiplier for the selected row (the current intensity values are multiplied by this number), and a divider for the selected column. Notice how the multiplied row seems noisy; this is because the calculated new intensity values are more than 255 (the data is limited to an unsigned 8 bit integer), therefore the new values wrap to lower levels.

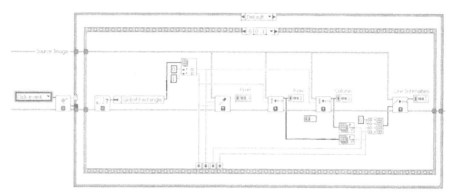

FIGURE 7.19 Grayscale image portions to mathematical values — wiring diagram.

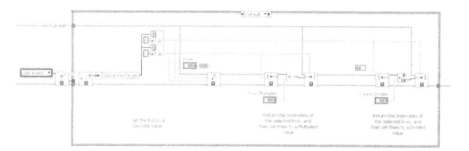

FIGURE 7.20 Mathematical values to grayscale image portions — wiring diagram.

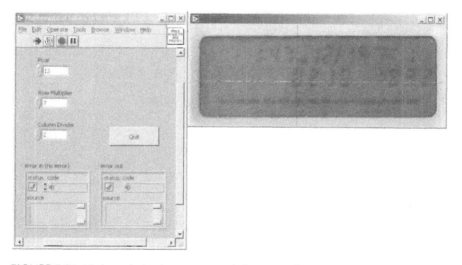

FIGURE 7.21 Mathematical values to grayscale image portions.

7.3.3 FILLING IMAGES

You can use IMAQ FillImage to change every pixel in an image to a particular intensity value. The basic example in Figure 7.22 was achieved using the code shown in Figure 7.23. Filling an image with one intensity value is analogous to using the Replace Array Subset primitive, changing every value in a two-dimensional array to a particular value (Figure 7.24).

Filling images with a particular intensity value may be useful to effectively destroy the source image's data, but IMAQ FillImage also has a mask input, allowing you to mask areas of the image so they are not filled. Consider the image of the power meter in Figure 7.25. If a mask is created that covers the barcode label, the rest of the image can be easily filled. Once the masking has been performed, only the necessary image data is retained. Filling the image (Figure 7.26) in this way also decreases the image's file size when using a compressing image file type (e.g., JPEG).

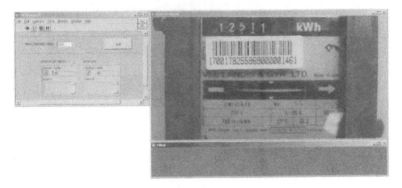

FIGURE 7.22 Fill image.

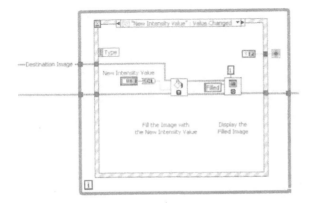

FIGURE 7.23 Fill image – wiring diagram.

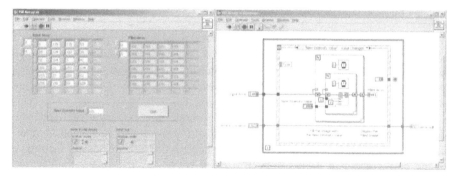

FIGURE 7.24 Fill Array.

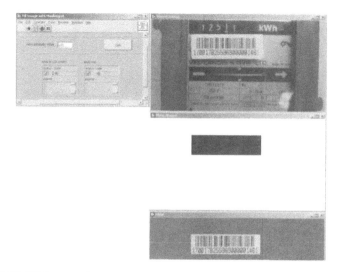

FIGURE 7.25 Fill image with masking.

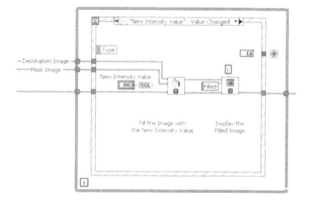

FIGURE 7.26 Fill image with masking — wiring diagram.

7.4 USER SOLUTION: IMAGE PROCESSING WITH *MATHEMATICA LINK FOR LABVIEW*

Dave Ritter
BetterVIEW Consulting
13531 19th Avenue
South Surrey, B.C., Canada V4A 6B3
Tel (604) 535 2456/Fax (604) 501 6102
e-mail: solutions@bettervi.com
Web site: http://www.bettervi.com

The IMAQ image processing tools are ideal for quickly developing a turnkey machine vision solution. However, in the early stages of a project, it is often convenient to acquire an image and experiment with it interactively. If you would like to experiment with different processing techniques, or perhaps optimize the lighting placement and camera position, an interactive and immediate command-line interface offers some compelling advantages.

One powerful, command-driven request/response interface well suited to interactive experimentation can be found in *Mathematica* from Wolfram Research. Mathematica's sophisticated symbolic capabilities and interactive "notebook" interface make it the tool of choice for countless scientists, researchers, and engineers worldwide, and with the help of BetterVIEW's *Mathematica Link for LabVIEW*, you can integrate Mathematica into your LabVIEW workflow. Not only does this solution offer you the power and immediacy of a command-line interface during the initial analysis and design phases, it also permits you to develop turnkey, Mathematica-enabled VIs with complete access to Mathematica's advanced symbolic, statistical, and mathematical capabilities.

Admittedly, this solution is not for everyone. However, if you are comfortable with technical image processing terminology, and some of the underlying mathematical principles, introducing Mathematica and the Mathematica Link for LabVIEW into your workflow will permit you a much greater degree of control over the low-level number-crunching details.

7.4.1 A TYPICAL LabVIEW/MATHEMATICA HYBRID WORKFLOW

In the initial phase, our objective is simply to acquire an image into Mathematica using LabVIEW and National Instruments hardware. Once the image has been acquired, we will process it interactively in the Mathematica notebook. When this initial experimentation phase is complete, we can determine whether to develop the final application using IMAQ, or if we would prefer the lower-level control offered by a hybrid LabVIEW/Mathematica solution.

Using Mathematica Link for LabVIEW, we begin by calling a LabVIEW-based video capture VI to acquire the image, as shown in the notebook in Figure 7.27. We have the option of either passing the image data across the "link" as numerical data, or simply saving the image in a file. In this example, we have chosen to save the image in a file to simplify the image data manipulations at both ends of the link.

(The image acquisition VI is configured to automatically save the acquired images into the Mathematica "Images" folder.)

Once the image has been acquired, it is loaded into Mathematica for interactive experimentation, as shown below. While this initially may look like a lot of typing to do, it is important to note that this notebook can be saved in it entirety. All of

```
<< LabVIEW'VIClient'                                                      ]

<< ImageProcessing'                                                       ]

link = ConnectToServer[LinkProtocol -> "PPC"]                             ]]

LinkObject[5555, 2, 2]                                                    ]]

inst = DeclareInstrument[
    "Macintosh HD.Desktop Folder.NI Week Demos.Build.ImageCapture.vi", {}, {}]

Instrument[Macintosh HD Desktop Folder NI Week Demos Build ImageCapture vi
    ImageCapture vi {} {}]                                                ]]

OpenVIRef[link, inst]                                                     ]]

ImageCapture vi loaded                                                    ]]

OpenPanel[link, inst]                                                     ]]

ImageCapture vi panel displayed                                          ]]

RunVI[link, inst]                                                         ]]

ImageCapture vi running                                                   ]]

ClosePanel[link, inst]                                                    ]]

ImageCapture vi panel closed                                             ]]

ServerShutdown[link]                                                      ]]

Server stopped                                                            ]]
```

FIGURE 7.27 Mathematica notebook.

the processing steps can be duplicated exactly simply by specifying a new file name and executing the notebook. As the notebook compiles a record of the entire session as you proceed, you can easily trace the path leading to the desired result. Furthermore, you can accurately evaluate and compare today's results with results from days, weeks, or even months past. All of the pertinent information is in the notebook, thus eliminating the need to keep accurate notes elsewhere — and much of the guesswork if others need to recreate your experiments (Figure 7.28).

It should be noted that the processing steps revealed in the figure take advantage of the Mathematica Digital Image Processing toolkit. However, because this toolkit is built entirely using native Mathematica functions, advanced users have the option of developing their own sophisticated image processing tools to meet very specific requirements, or to avoid incurring the extra expense of the toolkit.

7.4.2 TAKING THE NEXT STEP

If you are involved in research, the interactive experimentation phase outlined above may provide all the results you are seeking. In other situations, this initial phase will provide useful information if you are planning to process several images in a similar fashion. Perhaps you ultimately want to develop a stand-alone image processing solution to automate the process for less sophisticated end users. At this point, you can turn to the IMAQ tools and develop a standard LabVIEW-based solution, or you can develop a hybrid Mathematica Link for LabVIEW application.

```
<< ImageProcessing`

capture = ImageRead["capture2.bmp"];

Show[Graphics[capture]];
```

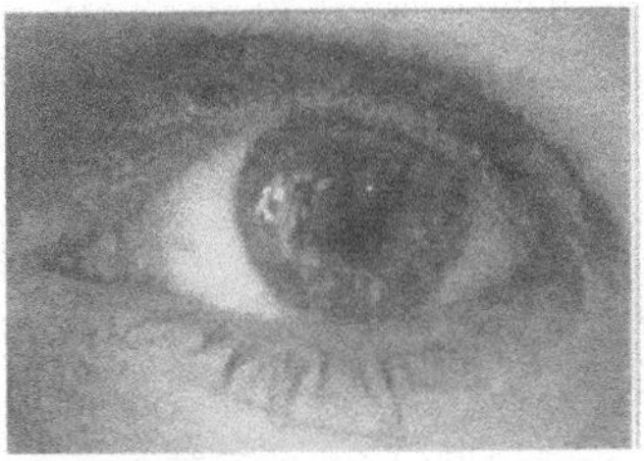

```
Show[Graphics[PlanarImageData[capture]]];
```

```
hist = ImageHistogram[capture];

Show[
  GraphicsArray[
   ListPlot[#, PlotJoined → True, PlotRange → {0, 0.04},
     Ticks → {{50, 100, 150, 200}, {0.02, 0.04}}, AspectRatio → 2/3,
     DisplayFunction → Identity] & /@ hist]];
```

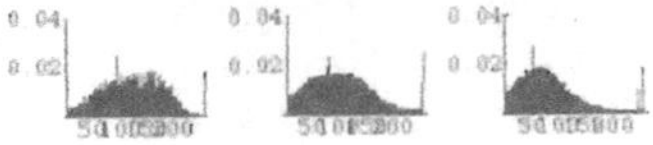

```
Show[GraphicsArray[
  Graphics /@ {Threshold[
    EdgeMagnitude[ToGrayLevel[capture], SobelFilter[]], 32],
    ZeroCrossing[
    EdgeMagnitude[ToGrayLevel[capture], LoGFilter[21, 21]]]}]];
```

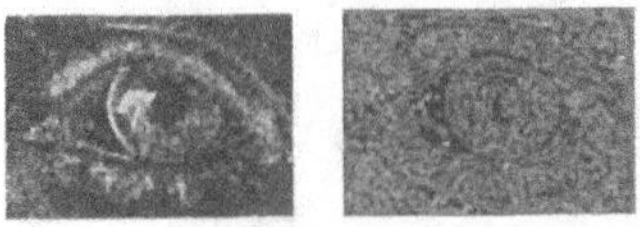

```
grayscl = ToGrayLevel[ScaleLinear[Log[2., 1. + capture], {0, 255}]];

regofint = Polygon[Reverse /@ Round[{{90, 70}, {90, 165}, {165, 70}, {165, 165}}]]

Polygon[{{70, 90}, {165, 90}, {70, 165}, {165, 165}}]

Show[Graphics[RegionProcessing[EdgeMagnitude[#1, SobelFilter[]] &, grayscl, regofint]]];
```

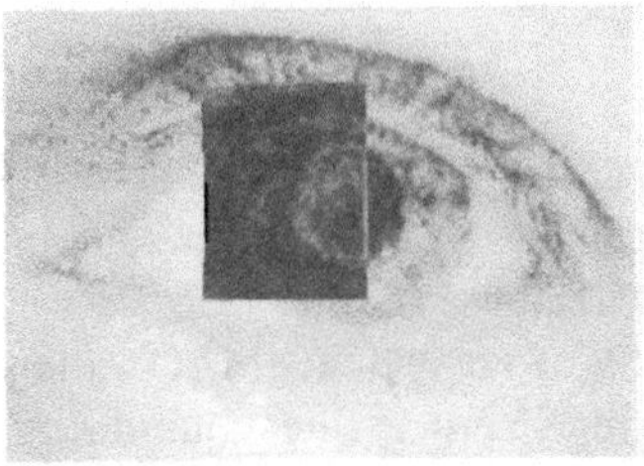

FIGURE 7.28 An interactive experimentation example.

Link-based hybrids take advantage of LabVIEW's easy-to-use GUI and Mathematica's advanced mathematical capabilities.

7.4.3 LabVIEW/Mathematica Hybrid Applications

A simple example of this hybrid approach is shown in Figure 7.29. This VI was prepared as a demonstration of image processing using Mathematica Link for LabVIEW for NI Week 2002.

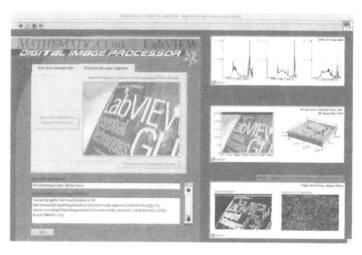

FIGURE 7.29 A hybrid Mathematica-LabVIEW application.

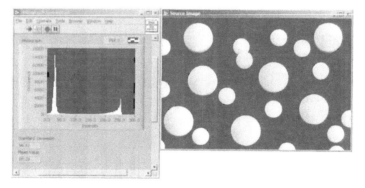

FIGURE 7.30 Histograph example.

Note that the only National Instruments component in this system is LabVIEW itself. The image acquisition camera was an inexpensive Firewire Webcam, and the image processing was accomplished using Mathematica and its Image Processing add-ons. When compared with the cost of NI vision hardware and the IMAQ toolkit, this demonstrates a lower-cost option for less-critical image processing requirements. More information about Mathematica can be found at http://www.wolfram.com. Mathematica Link for LabVIEW is available from Wolfram or directly from BetterVIEW at http://www.bettervi.com.

7.4.4 HISTOGRAMS AND HISTOGRAPHS

A histogram can be a very useful tool for determining the intensity distribution of an image. The formal definition of a histogram is

> ...a bar graph of a frequency distribution in which the widths of the bars are proportional to the classes into which the variable has been divided and the heights of the bars are proportional to the class frequencies.

A more practical explanation is to consider a histogram as a bar chart of intensity against intensity occurrence, with uniform intensity bandwidths (Figure 7.30).

The Vision Toolkit has two VIs that compute the histogram of an image: IMAQ Histograph and IMAQ Histogram. IMAQ Histograph returns a cluster containing the calculated histogram suitable for wiring directly into a front panel graph. It also returns the histogram's mean value and standard deviation (Figure 7.31).

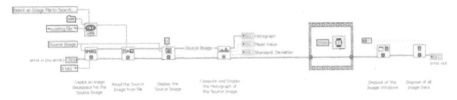

FIGURE 7.31 Histograph example — wiring diagram.

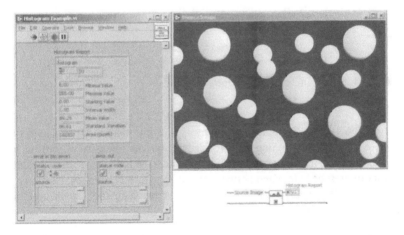

FIGURE 7.32 Histogram example.

IMAQ Histogram returns a cluster of data that includes the same information as IMAQ Histograph, but also contains the raw histogram as a one-dimensional array, the minimum and maximum values, and the histogram's area (in pixels) (Figure 7.32 and Figure 7.33).

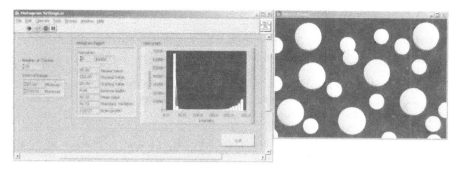

FIGURE 7.33 Histogram settings.

There is also the ability to define an *interval range* over which the histogram is to be calculated. Both `IMAQ Histogram` and `IMAQ Histograph` allow the user to specify the number of classes to sort the intensity values into, although the number of obtained classes differs from the specified value when the minimum and maximum boundaries are overshot in the *interval range* (Figure 7.34). The default number of

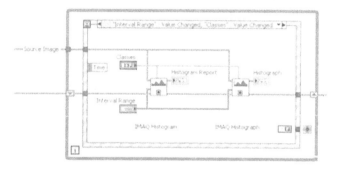

FIGURE 7.34 Histogram settings — wiring diagram.

classes is 256, which is most appropriate for 8-bit images. This value gives a complete class distribution: one class for each grayscale intensity in an 8-bit image. If the interval range minimum and maximum values are both set to zero (the default), the histogram will be calculated between 0 and 255 for 8 bit images. However, other image types will spread the number of classes between the lowest and highest detected pixel intensities.

7.5 USER SOLUTION: AUTOMATED INSTRUMENTATION FOR THE ASSESSMENT OF PERIPHERAL VASCULAR FUNCTION

Anne M. Brumfield and Aaron W. Schopper
National Institute of Occupational Safety and Health
1095 Willowdale Road
Morgantown, WV 26505-2888
Tel 304-285-6312/Fax 304-285-6265
Email: Znrl@cdc.gov

Methods for quantifying the injurious effects of vibration are a major focus for those evaluating hand arm vibration syndrome (HAVS) in the workplace. The physiologic changes in the microcirculation that occur following vibration exposure are not well understood; therefore, methods for evaluating blood flow in exposed worker populations are of particular interest.

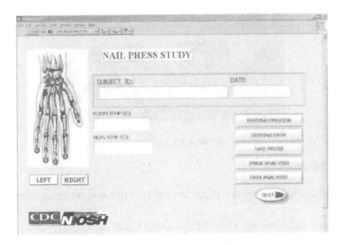

FIGURE 7.35 Initial user interface.

A nail press test (Figure 7.35) has been used by clinicians to assess peripheral circulatory function. Following a 10-second press of the fingernail, a visual determination of the rate of blood flow return is made, as evidenced by the return of color to the fingernail. Of a qualitative and subjective nature, the test has been deemed most meaningful when used as one among several in a battery of health effect assessment tests.[1,2] New instrumentation representing an automated version of the nail press has

[1] Olsen, N. (2002). Diagnostic aspects of vibration-induced white finger. *Int. Arch. Occup. Environ. Health*, 75: 6–13.
[2] Harada, N. (1987). Esthesiometry, nail compression and other function tests used in Japan for evaluating the hand–arm vibration syndrome. *Scand. J. Work Environ. Health*, 13: 330–333.

been built utilizing a CMOS camera, and force and photoplethysmograph transducers. A LabVIEW interface, enabling simultaneous image and data acquisition, provides a basis for the standardization of the force application and duration. The raw data is postprocessed and analyzed automatically, thereby eliminating observer subjectivity, and providing quantitative results with which worker populations can be evaluated.

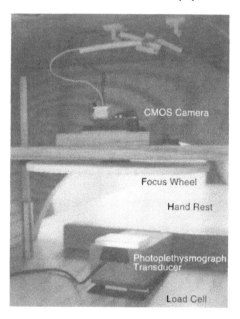

FIGURE 7.36 Image press prototype (rear view).

7.5.1 THE HARDWARE

The following components were incorporated into the system design (Figure 7.36):

- National Instruments AI-16XE-50 DAQ Card
- National Instruments PCI-1411
- National Instruments SC-2043-SG, strain gauge accessory
- Miniature CMOS Camera (1/3 inch format, 510 × 492 array size)
- Load cell
- Photoplethysmograph transducer, power supply, amplifier (Biopac)

7.5.2 THE SOFTWARE

7.5.2.1 DAQ

An interactive front panel (Figure 7.35) provides access to a variety of VIs within a waiting state machine. Relevant subject data is entered by the operator, including ID number, finger and room temperature, handedness and finger designation, which are incorporated into the appropriate file header or directory designation.

The `Resting Preview` VI provides image and transducer verification, and allows focus and finger placement adjustments to be made while familiarizing the subject with the procedure. The `Resting Data` VI collects 30 seconds of analog input data, providing baseline measurements of the pulse volume and resting finger force. From these measurements, the `Resting Mean` global variable is populated, and target force values for subsequent use during the nail press are computed from this value. As images are not acquired here, data is gathered at a high collection rate (1024 Hz) and provides an additional resource of data for any subsequently desired frequency analysis measurements requiring higher resolution.

The `Nail Press` VI utilizes its front panel interface to instruct and prompt the subject accordingly. The test consists of two sequences but starts only after the occurrence of a hardware trigger. This trigger starts a counter on the DAQ card, which generates a square pulse wave that serves to initiate each buffer of the image acquisition. During the initial sequence (the first 10 seconds), the interface provides an audible prompt and visual instructions to "Rest and Relax" while resting images and analog input data are gathered at 128 Hz. The subject is then instructed to "press and hold" in order to reach a desired target force, calculated by adding a load cell calibrated force to the RESTING MEAN global value. Once the target is exceeded, a countdown of 10 seconds begins, during which the subject maintains the target force level. The subject is then prompted to "rest and relax" for 20 additional seconds of data collection. A browser (Figure 7.37) is created which displays the images acquired during the press for verification of analysis results, and all images and raw data are written to file for postprocessing and analysis. The generation of time stamped data for timing verification purposes was essential, as this VI was expanded and elaborated upon for functionality. The use of software synchronization was not problematic at the low image acquisition rate (1 Hz); however, the use of hardware synchronization, which provides resolution in the order of nanoseconds, will be used when higher collection rates are desired.

7.5.2.2 Postprocessing and Analysis

Characteristic force and blood volume responses during the nail press are shown in Figure 7.38. The nail press sequence may be reviewed within the `Image Analysis` VI, which cycles through the acquired images at a user-defined speed (Figure 7.39). As each image is displayed, a histogram is generated and data from this report is unbundled and saved for further analysis. Although acquired as RGB images, the histograms are generated from their 8-bit representations. From the histogram data, selections are made for those classes demonstrating the most significant change during the press, and an evaluation of the pixel numbers within these bins over time generates a curve which, not surprisingly, resembles the curves of the force and blood volume results.

Critical to the automation of the nail press was the development of an algorithm that could be applied in the analysis of all of the data. Such an algorithm was developed and incorporated into both the `Image Analysis` and `Data Analysis` VIs. The typical image, force and blood volume data all revealed steep slopes during the nail press; therefore, the data were analyzed by taking step derivatives along the

 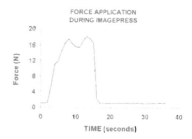

FIGURE 7.37 Changes during a nail press.

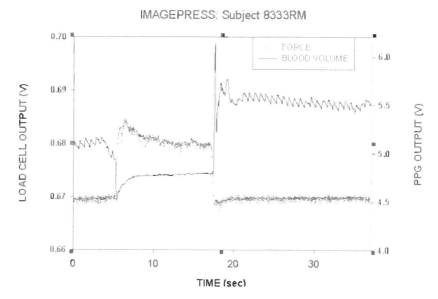

FIGURE 7.38 Representative force and blood volume data.

course of the press. Plotting the absolute values of these results reveals the areas of maximal change as defined by sharp peaks above baseline. The times corresponding to these peaks are calculated, and represent the times of maximal change during the nail press. The final results following the application of the step differentiation algorithm to the image data, and the force and blood volume data respectively, are shown in Figure 7.40.

7.5.3 SUMMARY

Using results gained from the developed instrumentation, it is possible to more accurately determine the times, which parallel the application of force and the return of blood flow and color to the finger. As demonstrated, the accuracy of the result is limited by the resolution of the acquisition system; therefore increasing the image acquisition rate above 1 Hz will certainly provide more precise determinations. This automated system provides a valuable method for the evaluation of blood flow in vibration-exposed worker populations.

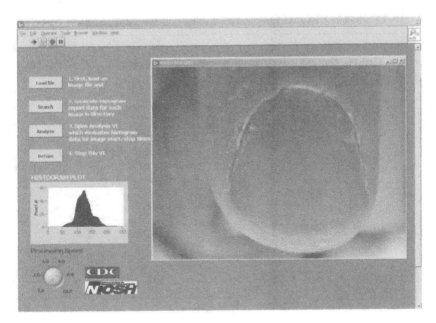

FIGURE 7.39 Pattern matching image analysis user interface.

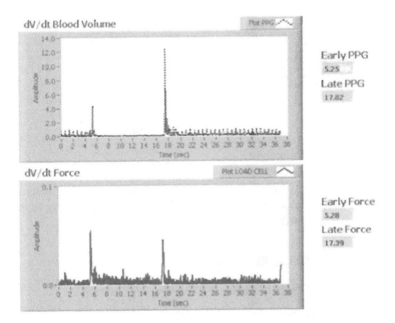

FIGURE 7.40 Time extraction — load cell and photoplethysmograph data.

7.6 INTENSITY PROFILES

Intensity profiles are one of the most commonly used vision tools. They allow the user to determine the pixel intensities and therefore their rates of change about a ROI (Figure 7.41).

The simplest intensity profile is the *point* — a single value that represents the intensity of the PUI. IMAQ Light Meter (Point) is used to return this value from a defined x,y position.

Pixel Contour

A *line* intensity profile is a one-dimensional array of values that represent the intensity of the pixels along the line, as drawn from the first to the last point. This example was achieved using IMAQ Light Meter (Line).

Line Contour

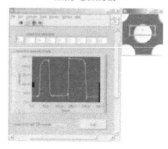

A closed ROI returns a one-dimensional array whose values trace around the ROI from the initial defined point in a clockwise direction. As shown in the example, the intensity profile midpoint is approximately 125 pixels of travel from the first ROI point, therefore the first 125 points of the profile correspond to the upper and right sides of the ROI. The values after 125 represent the lower and left sides of the ROI.

Closed ROI Contour

FIGURE 7.41

IMAQ ROIProfile uses the input image and a defined ROI to determine the intensity contour, which it returns in a format appropriate to wire directly to a front panel graph (Figure 7.42).

7.7 PARTICLE MEASUREMENTS

7.7.1 SIMPLE AUTOMATED OBJECT INSPECTION

One of the simplest machine vision VIs to begin toying with is IMAQ Count Objects. Although the name suggests that it will return a number of detected objects, it does much more than that. Consider an acquired image of some pharmaceutical tablets

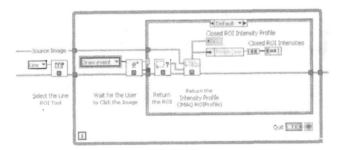

FIGURE 7.42 Closed ROI contour — wiring diagram.

(Figure 7.43). By using IMAQ Count Objects you can obviously count the number of tablets, but it also returns quite a lot of information about each of them, including each tablet's position with respect to the top left corner of the image, its bounding box, and how many holes (if any) were detected within it.

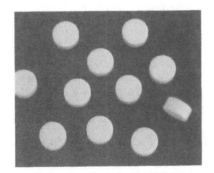

FIGURE 7.43 White tablets with a black background.

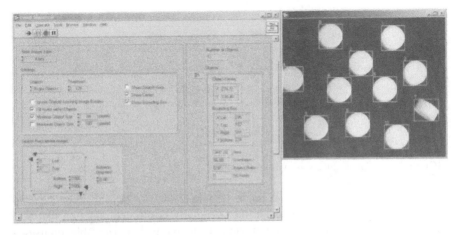

FIGURE 7.44 Count objects.

TABLE 7.4
"Count Objects Settings" Cluster Elements

Setting	Description
Objects	You can select whether the search routine should look for dark or light objects (with respect to the background).
Threshold	This is the grayscale intensity that is used as a threshold level.
Ignore Objects touching Image Borders	If a detected image intersects the defined search rectangle, it is not included as a counted image.
Fill Holes within Objects	If a detected object contains holes (regions within its bounds that are on the opposite side of the threshold descriptor) then they are filled, and the object is considered as a single entity.
Minimum Objects Size & Min Size	If an object is detected that is smaller than this size, it is ignored. This can be very useful when you do not have a perfectly populated image (e.g., excessive noise, dust or other contaminants are creating false "objects").
Maximum Object Size & Max Size	As you might expect, any detected objects larger than the Max Size are ignored.
Show Search Area	If set to true, a green box is overlaid on the image to show the search area.
Show Bounding Box	As demonstrated in the example above, the bounding box of detected objects is shown as an overlaid red box with an integer corresponding to the object's number (as referenced by the "objects" output array).
Show Object Center	Also as demonstrated above, the center of the Bounding Box is drawn as a small dot overlaid on the image.

This technique is very easily achieved by defining the settings of the search routine (Table 7.4). Although the search rectangle input suggests that by leaving it unwired it will search the whole image, this is not the case — you will need to set some sort of search rectangle either manually or programmatically for the IMAQ Count Objects routine to work (Figure 7.45).

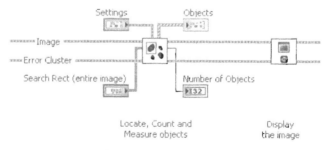

FIGURE 7.45 Count objects — wiring diagram.

As you can see from the example (Figure 7.44), 11 objects were detected (numerated from 0 to 10), including the irregularly positioned tablet on the right side of the image, and the partially obscured tablet on the left. The tablet on its end is detected as an object, because this routine is simply looking for objects that stand out from the background — not those that conform to particular predefined criteria, such as shape, color or size.

If the search rectangle is limited, IMAQ Count Objects will include objects partially bound by the search rectangle, and return their positions, bounding boxes, and number holes with respect to the image within the search rectangle (Figure 7.46).

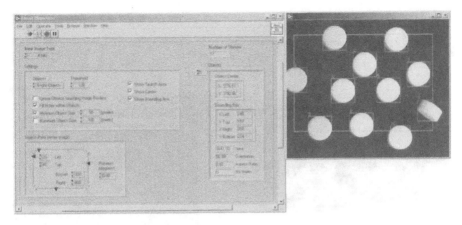

FIGURE 7.46 Count objects — limited search rectangle.

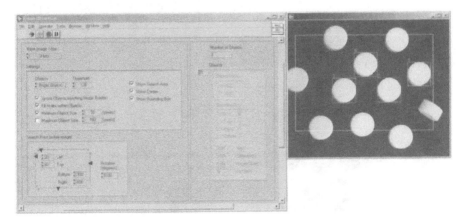

FIGURE 7.47 Count objects — objects touching search rectangle.

Setting the options to ignore objects that touch the search rectangle will consider only objects that are wholly within the rectangle (Figure 7.47).

7.7.2 MANUAL SPATIAL MEASUREMENTS

Measuring the distance between two points can be very easily accomplished by using Pythagoras' theorem (Figure 7.48). If we can determine the spatial coordinates of the two points between which we want to measure, then calculating the

$$Distance\left[\left(x_A, y_A\right), \left(x_B, y_B\right)\right] = \sqrt{\left(\Delta x\right)^2 + \left(\Delta y\right)^2}$$

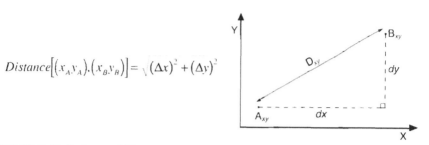

FIGURE 7.48 Pythagoras' Theorem.

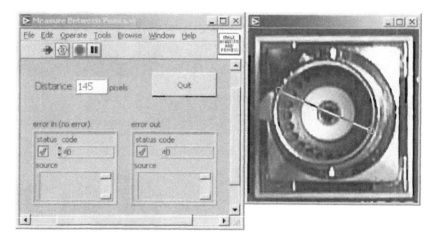

FIGURE 7.49 Measure between points.

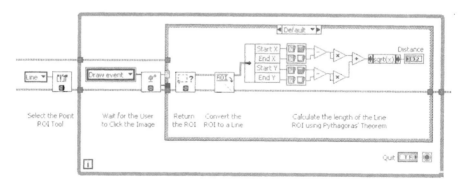

FIGURE 7.50 Measure between points — wiring diagram.

distance between them is trivial. For example, if a user draws a line ROI between two points, the pixel distance between them can be easily determined (Figure 7.49 and Figure 7.50).

While this is simple when we know the spatial positions of the two points, it becomes more difficult when their coordinates must be determined programmatically.

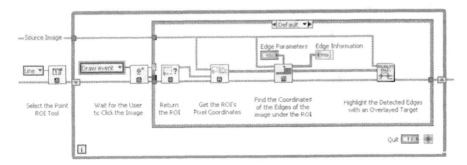

FIGURE 7.51 Finding edges and their parameters — wiring diagram.

7.7.3 FINDING EDGES PROGRAMMATICALLY

An edge in an image is generally defined as a feature interface with high contrast; if an image's intensity changes dramatically within a small spatial range, then it can be considered as an edge. The IMAQ Edge Tool can be used to detect and classify edges in an image by inspecting the pixel intensity values along a trace applied to the image. The example in Figure 7.51 obtains the pixel coordinates of points along a line ROI drawn by the user, and uses them to search for edges.

The pixel coordinates are obtained using IMAQ ROIProfile, and are wired into IMAQ Edge Tool. IMAQ Edge Tool outputs the number of edges detected, their coordinates and an array of clusters detailing comprehensive information regarding each of the edges (Figure 7.52).

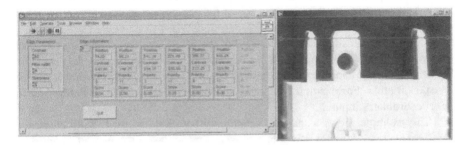

FIGURE 7.52 Finding edges and their parameters.

IMAQ Edge Tool is configured by defining the *edge parameters* to use during the search. *Contrast* specifies the threshold for the edge's intensity difference between features — edges with a contrast less than this threshold are not treated as edges, and are subsequently ignored. The contrast is calculated as the difference between the average pixel intensity before the edge and the average pixel intensity after the edge over the *filter width* (Figure 7.53), which is the maximum and minimum number of pixels over which you define an edge can occur. A small filter width ensures that the change in contrast occurs rapidly, whereas a larger filter width allows for a more gradual change.

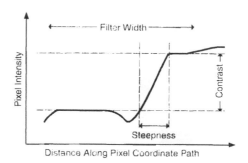

FIGURE 7.53 Edge parameters.

Consider a pixel position a. The contrast is determined by considering the pixels along the line of interest, both before and after a, that exist within the filter width:

$$Contrast = \left[\sum_{x=a-\left(\frac{f_w}{2}\right)}^{x=a} I(x) \right] \div \left[\sum_{x=a}^{x=a+\left(\frac{f_w}{2}\right)} I(x) \right]$$

The *edge information* array output of IMAQ Edge Tool contains an array of clusters, each describing respective detected edges. The two elements of the cluster describe the *position*, which defines the pixel coordinate location of the detected edge. This output is not the (x,y) position of the edge, nor does it represent the pixel number along the ROI where the edge is detected. When IMAQ Edge Tool executes, it searches the intensities that correspond to the pixel locations defined in the *pixel coordinates* input. This input in not necessarily linear, nor continuous. It is possible to define the pixel coordinates, so that IMAQ Edge Tool inspects pixels that are far from each other. For example, the array shown in Figure 7.54 is an entirely valid pixel coordinates input.

This example shows a spatial discontinuity between the 3rd and 4th elements. IMAQ Edge Tool does not interpolate between these coordinates, instead it considers them as continuous, and applies the edge detection across them.

The edge information array output also contains the *contrast* of the detected edge, and the *polarity*. A polarity of 1 indicates a rising edge (the intensities are changing from dark to light), whereas a polarity of −1 indicates a falling edge (light to dark).The final output cluster element is called *score*, and is unused.

Pixel Coordinates

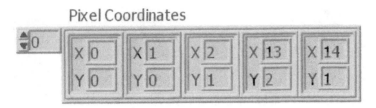

FIGURE 7.54 Pixel coordinates array.

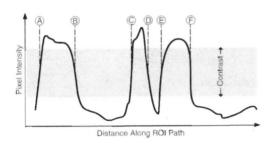

FIGURE 7.55 Example ROI path intensity plot.

7.7.4 USING THE CALIPER TOOL TO MEASURE FEATURES

IMAQ Edge Tool is very useful in finding edges along a ROI, but it is then up to the programmer to parse the returned information and attempt to understand what the edges mean with respect to the image. IMAQ Caliper Tool is able to detect and classify *edge pairs*, which are two consecutive edges with opposite polarities.

Consider the line ROI intensity plot in Figure 7.55. Edges exist (noted A through F), and can be classified using their polarities (A, C and E have a polarity of 1, whereas the polarities of B, D and F are –1). Using IMAQ Caliper Tool, the distances between the polarity pairs (AB, CD and EF) can be measured.

Consider the example in Figure 7.56. Not only have the obvious primary edge pairs been detected and measured, data concerning secondary pairs (pairs containing voids) has also been considered. Considering the ROI intensity plot in Figure 7.55, the primary pairs are AB, CD and EF, and the secondary pairs are AD, AF and CF.

Defining the *edge parameters* in the same way as IMAQ Edge Tool, the *caliper parameters* configures IMAQ Caliper Tool. The caliper parameters cluster contains three elements: polarity, separation and separation deviation (Table 7.5). The example in Figure 7.56 was achieved using the code shown in Figure 7.57.

After IMAQ Caliper Tool has detected and classified the edge pairs, their details are overlaid using Show Caliper Measurements.vi (available on the accompanying CD-ROM). This VI uses both the *caliper report* and the *edge coordinates* outputs from IMAQ Caliper Tool. The format of the edge coordinates array is different to that output by IMAQ Edge Tool — it does not represent simply a list of the detected edges, but each pair of array elements are the edge pairs detected. Considering the ROI intensity plot example in Figure 7.55, the coordinates contained

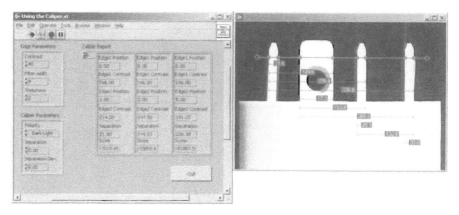

FIGURE 7.56 Using the caliper.

TABLE 7.5
"Caliper Parameters" Cluster Elements

Element	Description
Polarity	This controls what are to be considered the leading and trailing edges during the search. There are five possible polarity settings: None: All edges are considered as leading and trailing edges. Dark-Light: Leading=Dark, Trailing=Light Light-Dark: Leading=Light, Trailing=Dark Dark-Dark: Both Leading and Trailing=Dark Light-Light: Both Leading and Trailing=Light
Separation	Specifies the expected spatial difference between the edge pairs. Only edge pairs around this value (within the tolerance defined by the *separation deviation*) are considered. If *separation* is set to 0, all edge pairs detected are considered.
Separation deviation	Defines the tolerance value for the *separation* between the edges.

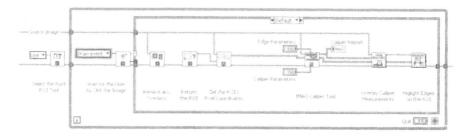

FIGURE 7.57 Using the caliper — wiring diagram.

in the array would represent the edge pairs in the following order: AB, AD, AF, CD, CF, EF, therefore the array element values would be as shown in Figure 7.58.

FIGURE 7.58 ROI intensity plot edge coordinates.

7.7.5 RAKING AN IMAGE FOR FEATURES

7.7.5.1 Parallel Raking

Just as a garden rake has several tines that create parallel lines in the dirt, an image rake is a series of parallel line ROIs, defined by their start and end points, and the relative distance between each line (Figure 7.59 and Figure 7.60).

Raking an image can be very useful when trying to find a feature in an image. Consider the example in Figure 7.59. If the part were shifted vertically, a single horizontal line ROI may fail to locate the hole in the pin, whereas using a rake can determine where the hole is, and its dimensions. IMAQ Rake is configured using the inputs in Table 7.6. The IMAQ Rake function returns an array of clusters, where each array element corresponds to a rake line. The clusters contain an array of clusters, each specifying the edge coordinates found along the rake line.

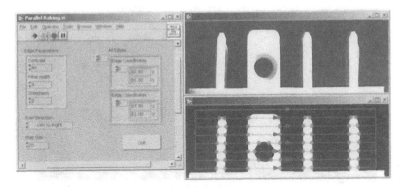

FIGURE 7.59 Parallel raking example.

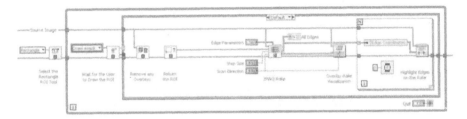

FIGURE 7.60 Parallel raking example — wiring diagram.

TABLE 7.6
"IMAQ Rake" Inputs

Input	Description
Edge Parameters	This input defines what should be considered as an edge. See Section 7.7.3, Finding Edges Programmatically, for more details.
Scan Direction	You can define the direction that edges are searched for along the lines. The following settings are permitted: Left to right (default) Right to left Top to bottom Bottom to top
Step Size	Defines the distance, in pixels, between the parallel lines inside the rectangular region.

7.7.5.2 Concentric Raking

Concentric circles share a common origin (as an aside, concentricity is the opposite to eccentricity), so concentric raking occurs when the rake lines follow circles with a common origin. Concentric raking is particularly useful when inspecting parts that have features intersecting with circular features (Figure 7.61).

FIGURE 7.61 Concentric raking example.

In the example (Figure 7.61), IMAQ Concentric Rake has found edges along concentric circular paths as defined by the annulus ROI. The positions of the detected edges are the output, and can be used to reconstruct a line model of the image, or detect where sections of the cross beams are missing. IMAQ Concentric Rake use is similar to IMAQ Rake, except the *scan direction* is defined as either clockwise or counterclockwise (default) (Figure 7.62).

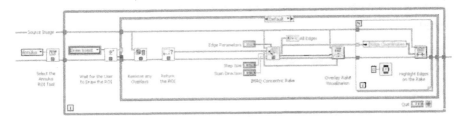

FIGURE 7.62 Concentric raking example — wiring diagram.

FIGURE 7.63 Spoke example.

7.7.5.3 Spoke Raking

When you are searching for circular features in an image, spoke raking can be a very powerful tool. Similar to parallel raking, spoke raking performs edge detection along rake lines, although they are not parallel. When spoke raking, the rake lines are portions of radial lines (Figure 7.63).

An arc or circular object can be mapped using the *edge coordinates* (Figure 7.64) output of IMAQ Spoke. IMAQ Spoke is configured using the inputs shown in Table 7.7.

TABLE 7.7

Input	Description
Edge Parameters	This input defines what should be considered as an edge. See section 7.7.3, Finding Edges Programmatically, for more details.
Scan Direction	You can define the direction that edges are searched for along the lines. The following settings are permitted: Outside to inside (default) Inside to outside
Step Size	Defines the distance, in pixels, between the radial lines inside the region.

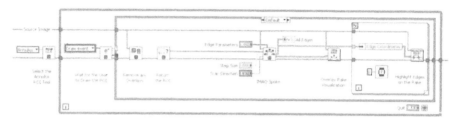

FIGURE 7.64 Spoke example — wiring diagram.

7.8 USER SOLUTION: SISSON-AMMONS VIDEO ANALYSIS (SAVA) OF CBFS

Bruce A. Ammons, Ph.D.
President — Ammons Engineering
8114 Flintlock Rd., Mt. Morris, MI 48458
Tel (810) 687-4288/Fax (810) 687-6202
e-mail: bruce@ammonsengineering.com
Web site: http://www.ammonsengineering.com
Joseph H. Sisson, M.D.
Professor of Medicine
University of Nebraska Medical Center

The University of Nebraska Medical Center needed a fast, efficient, and easy to use system to accurately measure the beat frequency of ciliated cells. Ammons Engineering designed the Sisson-Ammons Video Analysis (SAVA) system using LabVIEW and IMAQ Vision to record video directly into the computer and analyze the ciliary beat frequency (CBF), greatly reducing the time required to run each experiment.

7.8.1 INTRODUCTION

In the lungs, cells that line the trachea and bronchi (which are the conducting tubes of the lungs) have microscopic, hair-like appendages called cilia. Cilia beat in a synchronized pattern to propel mucus and inhaled particles out of the lungs. Dr. Joseph H. Sisson, Professor of Medicine at the University of Nebraska Medical Center, needed to upgrade a system developed using LabVIEW 2.1 in 1991 to a faster system using current technology. The system measures the CBF of airway tissues to determine the effect of various reagents, such as those found in medicines or in cigarette smoke. By adding the reagents to the cell samples and recording the change in the CBF over time, the effects of the medicines or toxins can be quantified. A typical experiment may have four or five different conditions with three or four samples for each condition. The system needed to be capable of recording each of the samples in half-hour intervals.

7.8.2 MEASUREMENT OF CILIA BEAT FREQUENCY (CBF)

To measure the cilia beat frequency; ciliated cells are grown on tissue culture dishes and incubated for several days until the cells are stable. This gives the cells time to group together and attach to the surface of the dishes. Before adding any reagents, a baseline CBF measurement is made for each sample. After adding reagents to the samples, brief measurements (approximately 5 seconds in duration) are recorded periodically for up to a 24-hour period. To measure the beat frequency, a sample is placed under an inverted phase contrast microscope and a group of motile cells is located, as shown in Figure 7.65. Video is acquired using a high-speed digital video camera mounted on the microscope that is capable of capturing digital images at up to 85 fps. At a single point, the light intensity alternates between light and dark each time cilia passes beneath the point. This variation in the intensity is used to measure the CBF. The operator selects a region of interest on the video image, which can be a point, a rectangle, or a line.

For each image in the video, the intensity of the pixels within the region is averaged and converted to a waveform representing the average intensity over time. This waveform is analyzed using standard LabVIEW tools to extract the dominant frequency and its amplitude.

The new system divides the CBF measurement process into two major phases. The first is the recording of the video, and the second phase consists of two alternative approaches to measuring the frequencies. Single point analysis allows the operator to choose the points of interest, and whole image analysis uses all of the points in the image.

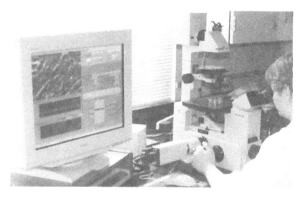

FIGURE 7.65 Locating cells using a microscope.

7.8.3 Video Recording

During the video recording phase, a live digitized video image is continuously displayed on the screen. The operator can adjust the frame rate, gain, and the number of images to store. From these settings, the maximum analysis frequency, spectrum frequency resolution, and video length in seconds are calculated and displayed. A timer displays the amount of time elapsed since the start of the experiment.

The video is acquired directly into the computer's memory using a circular buffer that contains a video history of the last few seconds. The size of the buffer is slightly larger than the number of images that will be recorded to disk. The video history enables the operator to quickly select various points on the video and instantly measure the CBF at each point. Once the operator has located an area in the sample with an adequate quantity of cells, they store the video to disk. The sample being analyzed is selected from the master list of samples, and the image sequence is stored along with the sample name, elapsed time and comments.

It is also possible to configure the system to record video at regular intervals without any operator interaction. The operator can leave one sample of interest on the microscope for recording over time. This way, the same cells are recorded at exact time intervals.

7.8.4 Single Point Analysis

After recording the samples, the video is analyzed to determine the beat frequencies (Figure 7.66). Analysis of recorded samples can be done during the longer intervals between recordings or after the experiment has been completed. The operator can select any previously recorded sample from a list, which includes a summary of analysis results for each sample. The recorded video is displayed in a repeating loop, along with information about the sample name, date, time, elapsed time, and comments as shown below. The video can be paused, single stepped forward or backward, and any image can be exported in a variety of formats for inclusion in reports.

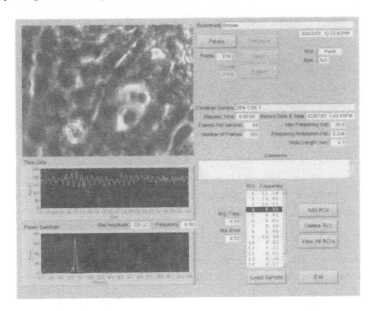

FIGURE 7.66 Single point analysis.

The same analysis tools available during the recording phase are available during the analysis phase, with a few extra features. The operator can store the results for multiple regions of interest, for which the mean and standard error are calculated. The results are automatically exported for graphing and inclusion in reports.

7.8.5 Whole Image Analysis

To significantly extend the capabilities of the system, a method to rapidly analyze large quantities of data was desired, and whole image analysis was the result. The image is divided into 4×4 pixel blocks, and each block is analyzed using the same routines as the single point analysis. The result of the analysis is displayed in two intensity graphs, one for frequency and one for amplitude, as shown in Figure 7.67. The operator can eliminate values outside the valid range. An intensity graph and a histogram of the remaining points are displayed, along with a statistical summary of the data.

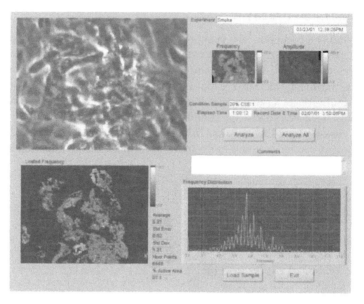

FIGURE 7.67 Whole image analysis.

Processing the entire image takes about 2 minutes for each sample. The procedure is fully automated for multiple samples, so the operator can start the analysis and go to lunch. When all of the samples have been analyzed, the results can be reviewed.

7.8.6 Design Challenges

One of the design challenges of this project was creating an efficient file structure for storing multiple experiments on the hard drive. Each experiment usually has a dozen or more samples, which results in hundreds of video segments per experiment. Another design challenge was selecting a method of compressing the video segments so that they load and save quickly without taking up too much space on the hard drive, yet retain enough detail to prevent errors in the analysis results.

7.8.7 Future Directions

Currently, the ciliated cells must be stored for several days so that they have time to attach to the tissue culture dish surface before the experiment begins. By adding the capability to track the position of slowly moving cells, it would be possible to measure the CBF of cells that are not attached to the dish surface. This would greatly reduce the amount of preparation time required for an experiment.

7.8.8 Results

Ammons Engineering developed a cost-effective solution based on a National Instruments hardware and software platform. The system is capable of handling multiple experiments simultaneously and has a simple, easy to use operator interface. Each sample can be analyzed in just a few minutes, drastically reducing the amount of time for recording and analyzing an experiment from 5 to 2 days or less.

7.9 ANALYTICAL GEOMETRY

The strict definition of analytic geometry is "the analysis of geometric structures and properties principally by algebraic operations on variables defined in terms of position coordinates." Analytical geometry can be used to determine the distance between points, the extrapolated intersection point of two non-parallel lines, fit circles and ellipses to points, and much more.

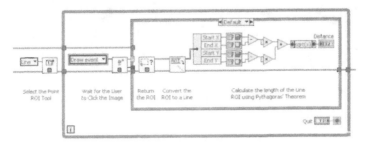

FIGURE 7.68 Manual measure between points — wiring diagram.

7.9.1 MEASURING BETWEEN POINTS

One can use the Pythagoras' theorem to determine the distance between two points:

$$Distance\left[\left(x_A y_A\right),\left(x_B y_B\right)\right] = \sqrt{\left(\Delta x\right)^2 + \left(\Delta y\right)^2}$$

This calculation can be performed using a VI similar to the one shown in Figure 7.68. Although this method is valid, the Vision Toolkit contains a VI that will calculate distances for you: IMAQ Point Distances (Figure 7.69 and Figure 7.70).

IMAQ Point Distances accepts an array input of clusters that describe points in an image (or any other mathematical space), and calculates the distance between each consecutive pair, which then can be overlaid on the source image, as shown in Figure 7.69.

7.9.2 LINES INTERSECTION

It is often useful to extrapolate two lines until they intersect, and then determine the intersection point and the angle between the lines.

Consider the example in Figure 7.71. IMAQ Lines Intersect has been used to extrapolate two line ROIs that have been drawn by the user (they could have also been dynamically created). Once the extrapolation is complete, the intersection point is determined, and the angle between the lines calculated. These results (the extrapolations, intersection and angle) have subsequently been overlaid to display the process (Figure 7.72).

The overlaying routine is completed within Show Lines Intersection.vi, available on the accompanying CD-ROM.

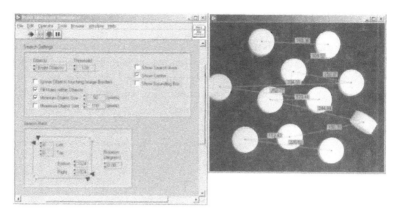

FIGURE 7.69 Point distances example.

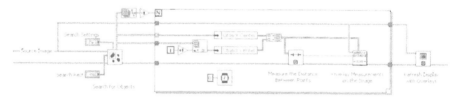

FIGURE 7.70 Point distances example — wiring diagram.

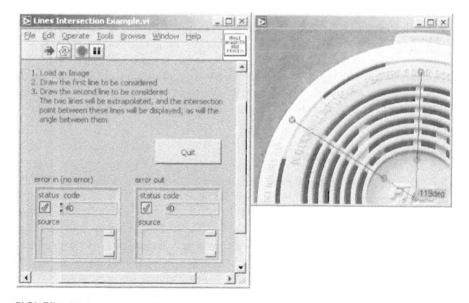

FIGURE 7.71 Lines intersection example.

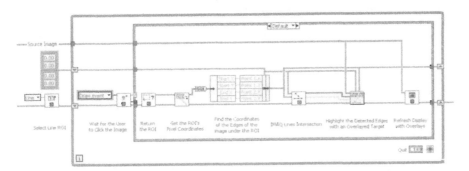

FIGURE 7.72 Lines intersection example — wiring diagram.

7.9.3 LINE FITTING

Fitting is a useful mathematical technique when used to discover trends in sample data sets, and can be just as valuable when examining images. Parts under inspection often contain regular repeating structures, and line fitting can be used as a quality control measure to ensure that parts are consistent. Consider Figure 7.73: A series of Australian flags has been printed on a sheet, although one of the flags has been printed with an incorrect alignment.

FIGURE 7.73 Fit line example.

Using IMAQ Count Objects and IMAQ Fit Line, it is obvious that the third flag from the left is not in the correct position. The Federation star (the seven-pointed star in the bottom left quadrant of the flag) has been detected, and their centers determined. This data has been then passed to IMAQ Fit Line, which determined the line of best fit. Features that are outside of a user-defined range of distances from the fitted line can then be regarded as inappropriate, and the part is failed (Figure 7.74).

IMAQ Fit Line does not perform a line of best fit in the true mathematical sense, as it can disregard some points that you include in the set to be considered. You are able to define a *pixel radius* that defines a corridor either side of the line of best fit, where only points within the corridor are considered. IMAQ Fit Line finds the line of best fit by first considering a subset of points that has been defined, and subsequently calculating the *mean square distance* (MSD) of this line (Figure 7.75).

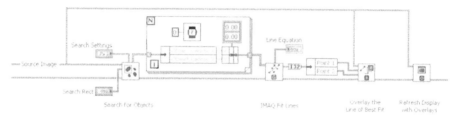

FIGURE 7.74 Fit line example — wiring diagram.

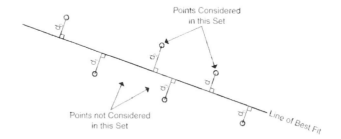

FIGURE 7.75 Line of best fit determination.

The mean square distance is calculated using the following equation:

$$MSD = \frac{\sum_{a=0}^{n-1} d_a^2}{n}$$

Once the MSD has been calculated, a new line of best fit is determined using a different set of points, and its MSD is also calculated. This process is repeated until all of the points in the input dataset have been considered (except those discarded for lying outside of the *pixel radius* range), and the set with the lowest MSD is returned as the line of best fit.

The MSD approach of line fitting was selected to minimize the impact of outlying points. Consider the parallel rake application shown in Figure 7.76. If the application requirement is to fit a line to the edge of the spanner head blade, the four extreme right detected edge coordinates are erroneous, although they are perfectly valid detected edges, they do not form part of the feature that we are looking for, and are ignored if the IMAQ Fit Line parameters are set appropriately.

7.9.4 CIRCLE AND ELLIPSE FITTING

If you are searching an image for a regular feature, it is often much faster to use a rake and reconstruct shapes from the detected edges rather than using a pattern match template. If you want either a circle or ellipse, the IMAQ Fit Circle and IMAQ Fit Ellipse analytical geometric tools can be helpful. IMAQ Fit Circle requires a minimum of three points to function, although extra valid points will increase its accuracy.

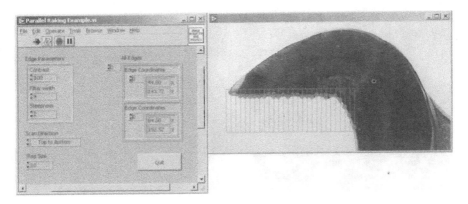

FIGURE 7.76 Wrench head blade parallel rake.

Consider the example in Figure 7.77. A parallel rake has detected edges that conform to specific edge parameters (contrast, filter width and steepness), which all lie on the inner and outer edges of a circle. Feeding the detected edges into IMAQ Fit Circle returns the center, radius, area and perimeter of the circle fitted to the edges. Spatially warping the image to create an oval displays how IMAQ Fit Circle still attempts to fit a circle to the detected edges (Figure 7.78)

IMAQ Fit Ellipse functions in a similar method to IMAQ Fit Circle, and returns the fitted ellipse's center position, the end points of both the major and minor axes, and the area. Inputting the minimum four points required into the IMAQ Fit Ellipse routine defines an ellipse, although using more points causes the routine to attempt to fit an ellipse.

FIGURE 7.77 Fit Circle example.

FIGURE 7.78 Skewed Fit Circle example.

8 Machine Vision

A generally accepted definition of machine vision is "...*the analysis of images to extract data for controlling a process or activity.*" Machine vision is a subdiscipline of artificial intelligence that uses video cameras or scanners to obtain information about a given environment, and to extract data from digital images about objects in the image. Machine vision takes an image in and outputs some level of description about the objects in it (i.e., size, position, correlation, etc.). Although applications for machine vision vary, there are generally only a few main categories:

- Quality assurance
- Sorting
- Material handling
- Robot guidance
- Calibration

Quality assurance represents probably the largest number of machine vision applications in use today. All that quality assurance machine vision systems do is inspect an image and decide whether the part should pass or fail, based on a predetermined set of rules. Although humans can be excellent quality assurance inspectors, vision systems are appropriate when images need to be inspected at high speed (often production lines require part inspection rates up to many complex parts per second), and for monotonous work. Vision systems do not become bored, or lack concentration; therefore, quality assurance can be much more highly controlled. Conversely, human inspectors can think and possibly expand or change the pass/fail rules — this can be both advantageous and disadvantageous. If the human finds a flaw in the product that is not part of the pass/fail criteria, then it can be added. On the contrary, humans can interpret rules incorrectly, often leading to quality assurance issues.

As its name suggests, *sorting* is the categorization of parts. Vision systems can detect part types based on predefined criteria and sort the parts accordingly. This detection can also be an extension of quality assurance, where pass and fail parts are segregated, and the production process continues without human interaction. Sorting application examples include the sizing of apples in a packing plant, the removal and remelting of excessively eccentric extruded copper tubes and the scrapping of fiber Bragg gratings that do not fit a bandwidth template.

Materials handling in hazardous environments can be simplified and automated using machine vision. Often, materials need to be handled quickly, and in areas that do not conform to desirable working conditions for humans. When considering the storage of products, warehouses populated by humans need to be well lit, and often use climate control to keep workers comfortable, which could be inappropriate for the product. Depending on the system used, machine vision systems can require minimal lighting or climate control to perform similar functions. Also, environments that are extremely dangerous for humans such as bomb disposal, toxic waste handling, high radiation and high temperature environments can be managed using machine vision process control.

Robot guidance often uses a close coupling of machine vision and robot control in industries such as arc welding for car construction, spray-painting and machine tool loading and unloading. Robot guidance routines can also encapsulate learning functions based on images acquired using machine vision.

Calibration of parts and assemblies can be automated using machine vision. Mechanical systems can be controlled, and their response monitored by a camera. A machine vision system can then analyze the part's response to the mechanical stimulus, and adjust the part to perform within a specified range. Companies that manufacture calculators, thermostats and keyboards often use machine vision systems to perform calibration.

National Instrument's Vision Toolkit implementation of machine vision comprises a varied tool palette, including functions that count and measure objects, distances and intensities; perform pattern matching; and read physical instrument displays.

Further information regarding machine vision theory can be found in the free online course *Introduction to Machine Vision* offered by Automated Vision Systems (http://www.autovis.com/courslst.html).

8.1 OPTICAL CHARACTER RECOGNITION

Optical character recognition (OCR) is the process of extracting textual data from an image, and is often used in the industrial world. From resolving the letters and numbers on a license plate of a moving car to scanning a document and saving the text to a file, OCR has become a widely accepted concept.

The IMAQ Vision OCR software is a toolkit purchased separately from the Vision Toolkit, and adds several very powerful character recognition routines that are able to quickly resolve text in a variety of ambient lighting conditions, fonts and printing qualities.

Although the toolkit is particularly suited to reading alphanumeric characters that have been produced using conventional typesetting and printing, it is also able to read common fonts used in automotive, electronics and pharmaceutical fields.

8.1.1 RECOGNITION CONFIGURATION

To perform a simple OCR, start by calling IMAQ OCR Recognition Config to set the recognition parameters (there are quite a lot of them, as shown in Table 8.1).

8.1.2 CORRECTION CONFIGURATION

After you have defined the recognition parameters, you can also invoke a correction routine called IMAQ OCR Correction Config. Using this VI will attempt to correct common textual errors based on whether you want to give preference to a particular character type.

TABLE 8.1
"IMAQ OCR Recognition Config" Parameters

Input	Description
Maximum point size	The OCR routine will only recognize text with a point size of this value or less (this value should be set to a value slightly above your expected point size).
Language	The OCR routine can recognize characters in English, French and Spanish.
Character set	If you know that the text you are trying to return has only uppercase letters, then you can instruct the OCR routine not to try and find other characters (lowercase letters, numbers, punctuation, etc). Limiting the character set will not only increase the speed of the OCR routine dramatically, but also inhibit common errors (for example, mistaking I for 1).
Character type	This input instructs the OCR routine to whether the expected text is normal, italic or a mixture of both (note: type checking is performed on a word-level, so mixing normal and italic characters within a word may cause the OCR routine to fail).
Recognition mode	There are four recognition modes: High Speed (Very fast, only good for low noise images with well spaced and defined characters). Nominal (Down-samples images to 200 dpi before scanning to improve speed — nominal and balanced are identical when source images that are already 200 dpi or less). Balanced (Good for a wide range of typeface variations and character qualities). Unoptimized (Disables all recognition optimizations).
Auto correction	When enabled, auto correction will cause the OCR routine to attempt to find the closest match to characters that it does not recognize at all. If disabled, unrecognized characters are ignored.
Output delimiter	This setting is only useful if you want to change the character that will appear between detected words. The available settings are None (disables the delimiter), Space (default), Tab, Comma and Full Stop ("Dot").

Several letters and numbers in the OCR's character library look similar:

Character	Similar Characters
The letter I	1 1 i ! | / \ [ ] '
The letter S	5 $
The letter O	0 Ø Ω o

Using these example characters, you can see how simple text can be incorrectly parsed:

Printed String	Possible OCR Results		
ISO 9001	ISO 9001	1SO 900I	!SO 900[
	I$Ø 9WW|	[50 900]	\$Ø 9ΩΩ/
	|SO 900|	iSo 9oo!	. . .

Using IMAQ OCR Correction Config, you can set the way ambiguously detected characters are interpreted. The VI has the inputs shown in Table 8.2.

TABLE 8.2
"IMAQ OCR Correction Config" Inputs

Input	Description
Character preference	This input depends on whether you want the OCR routines to favor alphabetic or numeric characters, or neither. For example, setting the character preference to *alphabetic* will prevent errors such as ISO ⇒ 1S0.
Method	The method input is similar to the recognition mode input for IMAQ OCR Recognition Config. Using this input, you are able to place more emphasis on whether the correction is faster or more accurate (default).
Aggressiveness	Aggressiveness defines to what extent the OCR search will attempt to parse a character that is ambiguous. When the OCR search detects a character that it is not sure about, it will compare it to a list of characters to find a possible match, and setting the aggressiveness changes the list of characters to be searched:
	Conservative List: 1I 1 | !] [0 O
	Aggressive List: 1 I1 | !] [0 O2 Z 5 S 8 B
	As you might expect, using the aggressive list will slow down OCR searching of ambiguous text.

8.1.3 OCR PROCESSING

Once you have set both the OCR recognition and correction parameters, all that remains to do is to perform the OCR search. There are two OCR searchVIs available in the Vision Toolkit: IMAQ Basic OCR and IMAQ Complex OCR. The only difference

TABLE 8.3

Input	Description
Optional rectangle	This input instructs the OCR search routine to only search a particular portion of the source image. If the user defines a region of interest at run time, then it can be advantageous to pass it to this input, as it can dramatically speed up the search process.
Rotation angle	If you know that the text you are looking for will be rotated at a particular angle, then it is a good idea to pass that value into the rotation angle input, otherwise the search engine may not detect the text.
Auto invert mode	By default, the OCR search engine will detect both dark text on a light background and light text on a dark background, but doing so requires two sequential searches. If you know that your text will be of one particular type, or you specifically want to ignore text of one of the types, you can set the invert mode using this input. Setting an invert mode will also increase the speed of the OCR processing.

between them is that IMAQ Complex OCR will output a report on each of the detected characters — the actual OCR search and recognition routines are identical.

When performing the OCR, you can define a few more parameters to make the search easier and faster. Both IMAQ Basic OCR and IMAQ Complex OCR have the inputs shown in Table 8.3.

8.1.4 PRACTICAL OCR

Figure 8.1 shows a VI that has performed an OCR on an image of a floppy disk. As you can see, the user has defined a region of interest (ROI) around some text on the source image, and the *OCR Text* output on the front panel of the VI has detected and resolved the text to read LabVIEW. This example was achieved using the OCR VIs as described previously, using the code shown in Figure 8.2.

Figure 8.3 shows an image with both alphabetic and numeric characters successfully detected.

8.1.5 DISTRIBUTING EXECUTABLES WITH EMBEDDED OCR COMPONENTS

If you intend to distribute your code as a stand-alone executable with OCR software embedded, you must copy the \OCR folder and its contents from \LabVIEW\Vi.lib\Vision\addons to the folder where the executable program will be located on any target machines, and copy Imaqocr.dll, Kitres.dll,

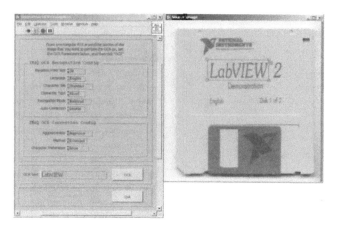

FIGURE 8.1 Interactive OCR example.

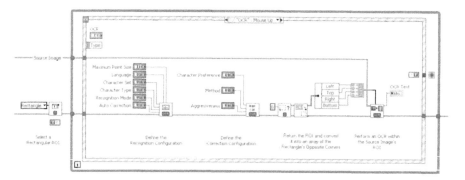

FIGURE 8.2 Interactive OCR example — wiring diagram.

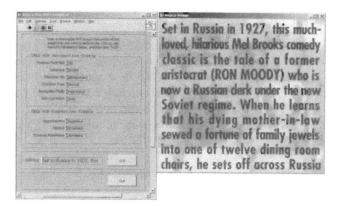

FIGURE 8.3 Mixed alphabetic and numeric text.

and Neurokit.dll to the Windows® system directory. Fortunately, this process can be automated with the use of commonly available software installation packages, including the LabVIEW Application Builder and SDS Software's *Setup2Go* (http://www.dev4pc.com). You are also required to purchase an OCR Software Run-Time License from National Instruments for each target PC.

8.2 PARSING HUMAN–MACHINE INFORMATION

You may have an experiment that uses physical displays that cannot be easily replaced with traditional data acquisition; therefore, acquiring an image of the physical display and using machine vision to interpret it into a data space can be useful. Machine vision is also helpful in the field of quality, where physical instruments can be controlled by a DAQ system, and their results monitored with image-based quality control software, or when the incidence of human error is high where operators enter information by hand.

8.2.1 7-Segment LCDs

Reading 7-segment displays (either LCD or LED based) is faster and simpler than optical character recognition as a much smaller number of characters need to be recognized — numerical digits only. The Vision Toolkit ships with VIs that read 7-segment displays, and are very easy to use. Resolving 7-segment displays takes place in two steps: learning and reading.

Because an LCD or LED display may contain more than one digit, the learning routines may need to detect several digits and return their respective ROIs; therefore, the first step is call IMAQ Get LCD ROI. Subsequently use IMAQ Read LCD to search for 7-segment digits within the generated ROIs. IMAQ Read LCD also has an output called segment status, which is an array of clusters that is populated with seven Boolean indicators arranged to simulate a 7-segment LCD digit. This output can be very useful if IMAQ Read LCD is not returning the values you expect, as it represents the segments that the VI could detect (Figure 8.4).

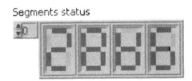

Segments status

FIGURE 8.4 The segment status array.

As you can see, the third digit is erroneous, so IMAQ Read LCD attempts to match the pattern with that of known digits. In this case, if we assume that the fault is that the upper segment has not been detected, IMAQ Read LCD will automatically correct the output from 28?6 to 2886.

IMAQ Read LCD is incredibly tolerant to poor-quality source images, and is able to overcome variable lighting across the image and noise issues. As long as the light drift (the difference between the average of the background pixel intensities in the top left and bottom right corners) is below 90, the contrast between the segments and background is at least 30, and each digit to be detected is larger than 18×12 pixels, IMAQ Read LCD is generally successful.

As described previously, using the Vision Toolkit LCD VIs is quite simple. Figure 8.5 is an example application. As you can see, the user has defined the ROI to encompass the two digits on the right, and IMAQ Get LCD ROI has detected two digits within the ROI. IMAQ Read LCD has then successfully determined that the two digits are both eights. This example was achieved using the code shown in Figure 8.6.

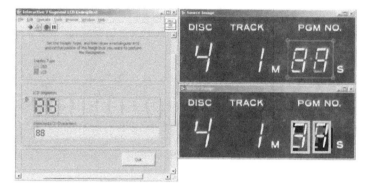

FIGURE 8.5 Interactive 7-segment LCD example.

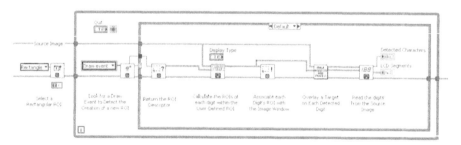

FIGURE 8.6 Interactive 7 segment LCD example — wiring diagram.

8.2.2 BARCODES

Labeling parts with machine-readable codes has become extremely common in recent years. From every packaged product in a supermarket, to library books, and even employee cards, barcodes are one of the cheapest and easiest systems that we can introduce. LED or laser handheld barcode scanners are inexpensive and are simple to interface to computers either through the PC's serial port or in line with the output of the keyboard (often known as a *keyboard wedge*). The Vision Toolkit includes VIs that can resolve and read barcodes that are part of an image. If you are already acquiring an image of an object for other vision-based reasons, it makes good sense to add barcode-reading code to limit the influence of human error when tracking work in progress (WIP).

8.2.2.1 One-Dimensional

The standard one-dimensional barcodes are the ones we are most accustomed to: a row of alternating black and white bars of varying widths representing the code. An example one-dimensional barcode is shown in Figure 8.7.

Reading one-dimensional barcodes with LabVIEW is achieved by using only one VI: IMAQ Read Barcode. Although most barcodes look quite similar, there are many different standards currently in use. IMAQ Read Barcode can read the following types:[1]

- Codabar
- Code 39: often referred to *Code 3 of 9*
- Interleaved 2 of 5
- Code 93
- Code 128: the most commonly used standard
- EAN 8
- EAN 13
- MSI
- UPCA

FIGURE 8.7 An example one-dimensional barcode.

Figure 8.8 is a simple barcode reader example. This example is based on the code in Figure 8.9. As you can see, IMAQ Read Barcode returns a cluster output called *extra info*. This cluster contains the elements shown in Table 8.4.

8.2.2.2 Two-Dimensional

As the name suggests, two-dimensional barcodes allow you to represent data in a two-dimensional matrix printed on a page. *Data Matrix* is a very common and area efficient two-dimensional barcode symbology that uses a unique perimeter pattern to help the barcode scanner determine each of the barcode's cell locations (Figure 8.10).

---

[1] The first three barcode types contain validation characters (similar to cyclic redundancy check error characters), which can be subsequently used to validate the barcode when using the Vision Toolkit VIs. The other barcode types either have no such validation codes embedded in them, or any validation is performed automatically because the barcode type requires the parsing routine to perform the decoding.

The Data Matrix standard can encode letters, numbers and raw bytes of data, including extended characters and Unicode. Large amounts of text and data can be stored when using two-dimensional barcodes: more than 800 ASCII characters can be encoded in a single barcode. For example, encoding the text `Barcode12345678` yields the results shown in Table 8.5 when using different barcode types.

FIGURE 8.8 Interactive barcode reader example.

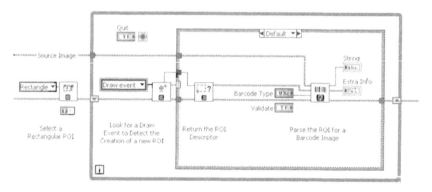

FIGURE 8.9 Interactive barcode reader example — wiring diagram.

The National Instruments Vision Toolkit does not offer support for two-dimensional barcodes by default, but Ammons Engineering has developed a toolkit for reading Data Matrix ECC200 two-dimensional barcodes from images in LabVIEW. This toolkit's only required parameters are the dimensions of the barcode, and is very fast (typical read times are under 100 ms). More information on the Ammon's two-dimensional barcode toolkit can be found at their Web site (http://www.ammonsengineering.com).

Dynamically creating two-dimensional barcodes using LabVIEW is also not supported by default, although ID Automation has an excellent ActiveX barcode creation toolkit available on their Web site (http://www.idautomation.com/activex).

TABLE 8.4
"IMAQ Read Barcode–Extra Info" Cluster Elements

Cluster Element	Description
Special character 1	Particular barcode types contain special leading and trailing characters, as defined below (if special characters do not exist, these values are returned as 0): • Codabar: start character • Code 128: function character (FNC) • EAN 8: first country code • EAN 13: first country code
Special character 2	This variable is similar to *special character 1*: • Codabar: stop character • EAN 8: second country code • EAN 13: second country code • UPCA: system number
Confidence level	When IMAQ Read Barcode executes, it measures the relative widths of the dark and light bars in the barcode, and the *confidence level* reflects the level of error in these width measurements. *Confidence level* returns values between 0 (poor confidence) and 1000 (high confidence); generally confidence levels over 700 are acceptable.

FIGURE 8.10 An example of a two-dimensional barcode.

8.2.3 INSTRUMENT CLUSTERS

Soliton Automation has developed a Vision Toolkit-based inspection system that inspects automotive instrument clusters for color, illumination and defects. The PC-based machine vision system uses National Instruments' image acquisition hardware to capture images, and customized application software inspects the quality of illumination, colors, sequences and currents. More information about this solution and methods of including similar code into your own application can be found at Soliton Automation's Web site (http://www.solitonautomation.com).

Barcode	Barcode Type
	Code 39 (3 of 9)
	Code 128
	Data Matrix

FIGURE 8.11

Glossary

1-Bit Image An image comprising pixels that contain only a single bit of information — each pixel is either on or off. Usually, "on" refers to white, while "off" is black.

4-Bit Image An image file format that allows for 4 bits of image-based data per pixel. Such an image can contain up to 16 different colors or levels of gray.

8-Bit Image An image where each pixel contains 8 bits of information. An 8-bit pixel can take on one of 256 possible values. There are two common types of 8-bit images: grayscale and indexed color. In grayscale images, each pixel is represented as one of 256 shades linearly distributed of gray from 0 (usually black) to 256 (usually white), and therefore does not require a palette (but often contains one anyway). Indexed color images always contain a palette, and each pixel is an index to a color in the palette of 256 colors.

24-Bit Image A 24-bit image contains pixels made up of RGB byte triplets.

Additive Primary Colors Red, green and blue; the colors used to create all other colors when direct or transmitted light is used. They are called "additive primaries" because they produce white when superimposed.

Antialiasing A method of filling data that has been missed due to undersampling. Antialiasing is often used to remove jaggies by interpolating values between pixels, reducing the stairstepping artifact often found in high-contrast digital images.

AOI Area of interest; a rectangle within an image defined as two points within the image, with each side parallel with one of the image's axes. See also *ROI*.

Aspect Ratio The proportion of an image's size given in terms of the horizontal length vs. the vertical height. An aspect ratio of 4:3 indicates that the image is 4/3 times as wide as it is high.

Bezier Curve A curve created from endpoints and two or more control points that serve as positions for the shape of the curve. Originated by P. Bezier.

Binary A type of image in which the pixel values represent either light (1) or dark (0), with no intermediate shades of color.

Bit Block Transfer A raster operation that moves a block of bits representing some portion of an image or scene from one location in the frame buffer to another.

Bitmap An image that contains a value for each of its pixels. See also *Vector Image*.

Bit Plane A theoretical two-dimensional plane containing a single bit of memory for each pixel in an image.

Bounding Rectangle The smallest rectangle that fits around a given object, often rotationally restricted to be parallel to both image axes.

BMP Microsoft bitmap image file format extension.

Camera An image sensing device where an image is formed on a plane, and the energy of the image is sensed point by point.

Cartesian Coordinates An equally spaced grid that uniquely assigns every point in the space as a coordinate pair (two-dimensional: x, y) or triplet (three-dimensional: x, y, z). In imaging, each point is usually referred to as a *pixel*. Most images use the top-left as their origin (0,0). See also *Coordinates*.

Chroma-key An image blending function which replaces pixels of a specified hue range with pixels from a secondary image. This is often referred to as the *weatherman effect*, as weather forecasters often stand in front of a solid blue or green background, which is transformed into a weather map.

CIE *Commission Internationale de l'Eclairage*. The International Commission of Illumination. The CIE is a standards organization that provides specifications for the description of device-independent color.

CMY/CMYK Cyan, magenta, yellow, black. Printers coat paper with colored pigments that remove specific colors from the illumination light. CMY is the subtractive color model that corresponds to the additive RGB model. Cyan, magenta and yellow are the color complements of red, green and blue. Due to the difficulties of manufacturing pigments that will produce black when mixed together, a separate black ink is often used and is referred to as K (The symbol B cannot be used, as it represents blue).

Color Model/Space A mathematical coordinate system used to assign numerical values to colors. There are many ways to define such spaces, including CMY, CMYK, IHS, HSL, HSV and RGB.

Compression An image processing method of saving disk and memory requirements by reducing the amount of space required to save a digital image. Image data is rewritten so that it is represented by a smaller set of data. See also *Lossless* and *Lossy* Compression.

Compression Ratio The ratio of a file's uncompressed size and its compressed size.

Convolution An image processing operation used to spatially filter an image, defined as a kernel. The size of the kernel, the numbers within it, and often a single normalizer value define the operation that will be applied to the image. The kernel is applied to the image by placing the kernel over the image to be convolved and moving it through the image so that it is individually centered over every pixel in the original image. See also *Kernel*.

Crop An image processing method of removing a region near the edge of the image, retaining a central area.

Decompression Returning an image from a compressed format. See also *Compression*.

Dilation See *Morphology*.

Dithering The method of using neighborhoods of display pixels to represent one image intensity or color. This method allows low intensity resolution display devices to simulate higher resolution images. For example, a binary display device can use block patterns to simulate grayscale images. See also *Halftone*.

DLL Dynamic linked library. A compiled and linked collection of computer functions that are not directly bound to an executable. These libraries are linked at run time by the operating system. Each executable can link to a commonly shared DLL, saving memory by avoiding redundant functions from coexisting.

DPI Dots per inch. The number of printer dots that can be printed or displayed along a 1 in., one-dimensional line. Higher DPI values represent higher resolutions.

Edge A region of contrast or color change. Edges are often useful in machine vision because optical edges often mark the boundary of physical objects.

Edge Detection A method of isolating and locating an optical edge in a given digital image. See also *Edge*.

Encoding The manner in which data is stored when uncompressed (binary, ASCII, etc.), how it is packed (e.g., 4-bit pixels may be packed at a rate of 2 pixels per byte), and the unique set of symbols used to represent the range of data items.

EPS Adobe encapsulated postscript file format extension.

Equalize An image processing algorithm that redistributes the frequency of image pixel values so that any given continuous range of values is equally represented.

Erosion See *Morphology*.

File Format A specification for encoding data in a disk file. The format dictates what information is present in the file and how it is organized within it.

Filter An image processing transform that removes a specified quantity from an image.

Frame A single picture, usually taken from a collection of images such as in a movie or video stream.

Frame Buffer A computer peripheral that is dedicated to storing digital images.

Gain and Level Gain and level are image processing terms that roughly correspond to the brightness (gain) and contrast (level) control on a television. By altering the gain, the entire range of pixel values are linearly shifted brighter or darker, whereas level linearly stretches or shrinks the intensity range.

Gamma Correction A nonlinear function that is used to correct the inherent nonlinearity of cameras and monitors.

Geometric Transform A class of image processing transforms that alter the location of pixels.

GIF CompuServe Graphics Interchange File format extension. Uses the LZW compression created by Unisys, which requires special licensing. All GIF files have a palette.

GUI Graphical user interface. A computer–user interface that substitutes graphical objects for code to enhance user interaction.

Gray Level A shade of gray assigned to a pixel.

Grayscale A type of image in which the pixel values are proportional to the light intensity corresponding to a small spot in the image. Zero is usually black and higher numbers indicate brighter pixels.

Halftone The reproduction of a continuous-tone image on a device that does not directly support continuous output. This is done by displaying or printing pattern of small dots that from a distance can simulate the desired output color or intensity.

Histogram A tabulation of pixel value populations usually displayed as a bar chart where the x axis represents all the possible pixel values and the y axis is the total image count of each given pixel value. Each histogram intensity value or range of values is called a bin. Each bin contains a positive number that represents the number of pixels in the image that fall within the bin's range. When the collection of bins is charted, the graph displays the intensity distributions of all the images pixels.

HSL Hue, saturation and lightness. A method of describing colors as a triplet of real values. The hue represents the wavelength of the color, the saturation is the depth, and the lightness refers to how black or white a color is.

Huffman Encoding A method of encoding symbols that varies the length of the code in proportion to its information content. Groups of pixels that appear frequently in an image are coded with fewer bits than those of lower occurrence.

IHS Intensity, hue and saturation.

Image Format Refers to the specification under which an image has been saved to disk or in which it resides in computer memory. There are many commonly used digital image formats in use, including TIFF, BMP, GIF, and JPEG. The image format specification dictates what image information is present and how it is organized in memory. Many formats support various subformats.

Image Processing The general term refers to a computer discipline wherein digital images are the main data object. This type of processing can be broken down into several subcategories, including compression, image enhancement, image filtering, image distortion, image display and coloring and image editing. See also *Machine Vision*.

Intensity Graph A native LabVIEW front panel indicator, which can display 256 colors (selected, via a color table, from the available 16 million). The value that is input to each pixel determines both the color and the intensity that is displayed.

Jaggies A term used to describe the visual appearance of lines and shapes in raster pictures that results from using a drawing grid of insufficient spatial resolution.

JPEG Joint Photographic Experts Group image compression capable of both lossy and lossless compression.

JPG Joint Photographic Experts Group file format extension. See *JPEG*.

Kernel A matrix of pixels that is used as an operator during an image convolution. The kernel is set prior to the convolution in a fashion that will emphasize a particular feature of the image. Kernels are often used as spatial filters, each one tuned to a specific spatial frequency that the convolution is intended to highlight. See also *Convolution*.

LZW Lempel-Zev-Welch. An image compression method with lossless performance.

Level See *Gain and Level*.

LUT Look up table. A continuous block of computer memory that is initialized in such a fashion that it can be used to compute the values of a function of one variable. The LUT is set up so that the functions variable is used as an address or offset into the memory block. The value that resides at this memory location becomes the functions output. Because the LUT values need only be initialized once, they are useful for image processing because of their inherent high speed. See also *Palette*.

Lossless Compression A method of image compression where there is no loss in quality when the image is subsequently uncompressed (the uncompressed image is identical to its original). Lossless compression usually results in a lower in compression ratio than lossy compression.

Lossy Compression A method of image compression where some image quality is sacrificed in exchange for higher compression ratios. The amount of quality degradation depends on the compression algorithm used, and its settings.

Machine Vision A subdiscipline of artificial intelligence that uses video cameras or scanners to obtain information about a given environment. Machine vision processes extract information from digital images about objects in the image. Machine vision takes an image in and outputs some level of description about the objects in it (i.e., color, size, brightness). See also *Image Processing*.

Morphing An imaging process where one image is gradually transformed into a second image, where both images previously exist. The result is a sequence of in-between images which when played sequentially give the appearance of the starting image being transformed to the second image. Morphing is made up of a collection of image processing algorithms. The two major groups are warps and blends.

Morphology A neighborhood image processing algorithm similar to image convolution except that Boolean logic is applied instead of arithmetic. There are two types of morphology: binary and grayscale. The four major morphology operations are:
- Erosion: a filter which tends to make bright objects smaller.
- Dilation: a filter which tends to make bright objects larger.
- Opening: an erosion followed by a dilation.
- Closing: a dilation followed by an erosion.

Neighborhood Process A class of image processing routines that works on neighborhoods of pixels at a time. Each pixel in the new image is computed as a function of the neighborhood of the pixel from the original pixel. See also *Point Process*.

MPEG Moving Pictures Experts Group image compression.

MUTEX Mutual exclusion object. A program object that allows multiple program threads to share the same resource, such as file access, but not simultaneously. When a program is started, a MUTEX is created with a unique name, and then any thread that needs the resource must lock the MUTEX from other threads while it is using the resource. The MUTEX is set to unlock when the data is no longer required or the routine is finished.

Overlay An object, image or subimage that can be placed over a given image. The pixels from the original image are not altered, but the overlay can be viewed as if they had been. Usually used to place temporary text and annotation marks such as arrows and company logos over an image.

Packed Bits When a binary image is usually stored in computer memory 8 pixels per byte. Using this technique saves memory, but makes reading and writing any individual pixel somewhat more difficult, as most PCs cannot directly access memory in chunks smaller than a byte.

Palette A digital images palette is a collection of 3 LUTs, which are used to define a given pixels display color; one LUT for red, one for green and one for blue. A palette image requires its palette in order to be displayed in a fashion that makes sense to the viewer (without a palette describing what color each pixel is to be displayed with, images are most likely displayed as randomly selected noise). A grayscale palette occurs when each of the 3 LUTs are linear, as since each color component (R, G, B) will be an equal value, any pixels input to them will be displayed in a varying shade of gray. See also *LUT*.

Pattern Recognition/Recognition A subdiscipline of machine vision where images are searched for specific patterns. Optical character recognition (OCR) is one type of pattern recognition, where images are searched for the letters of the alphabet.

PCX ZSoft Corporation picture file format extension.

Picture Indicator A native LabVIEW front panel indicator that can display 32-bit color, using 8 bits for each color, R, G and B, with the left-most byte (α) unused. The intensity of each pixel in the picture control's colored image is proportional (for a linear gamma curve) to the sum of the RGB values. The color of each pixel is determined by the ratio of the RBG values for that pixel.

Pixel Picture element. The most fundamental element of a bitmap image.

Point Process A class of image processing that mathematically transforms each pixel individually, with no weighting from neighboring pixel intensities.

Posterize A special effect that decreases the number of colors or grayscales in an image.

Pseudocolor A method of assigning color to ranges of a grayscale image's pixel values. Most often used to highlight subtle contrast gradients or for visually quantifying pixel values. The applied color usually has no correspondence to the original, physical imaged scene.

Render The process of displaying an image.

Resolution There are two types of resolution in digital images: spatial and intensity. Spatial resolution is the number of pixels per unit of length along the x and y axis, whereas intensity resolution is the number of quantized brightness levels that a pixel can have.

RGB Red, green, blue. A triplet of numeric values used to describe a color.

ROI Region of interest. A specification structure that allows for the definition of arbitrarily shaped regions within a given image. A ROI is a placeholder that remembers a location within an image. The ROI can encompass the entire image. See also *AOI*.

Scan line See Raster.

Screen Coordinates Screen coordinates are those of the actual graphics display controller, with its origin at the upper lefthand corner of the display. See also *Coordinates*.

Segment A contiguous section of a raster line defined in physical coordinates by the triplet of its left-most point and length (x, y, l).

Systems Integrator A service provider who can deliver a custom hardware and software system, integrating various components from different manufacturers, often writing custom software and fabricating custom hardware.

TGA TrueVision Corporation file format extension.

TIFF Tagged Image File Format extension.

Thumbnail A small copy of an image usually used to display many images on the screen at once in a browser configuration.

Transform An algorithm that alters an image, often written as *xform*. See also *Point Transform, Neighborhood Transform* and *Geometric Transform*.

Triplet Three numbers which when used together represent a single quantity or location (e.g., RGB or x, y, z.)

Umbras Portions of a shadow that are neither totally light nor totally dark.

Video Signal A time varying signal where the amplitude of the signal is proportional to the light sensed on a point in the image, and the elapsed time identifies the location of the sensed point in an image.

Video Stream A sequence of still images that are transmitted and displayed in synchronous order given the appearance of live motion.

WMF Microsoft Corporation Windows MetaFile Format file format extension.

WPG Word Perfect Corporation file format extension.

Bibliography

TEXTS

IMAQ Vision Concepts Manual, 2002, National Instruments

LabVIEW Advanced Programming Techniques, Bitter, Mohiuddin, and Nawrocki, 2001, CRC Press LLC

LabVIEW Graphical Programming (2nd ed.), Gary W. Johnson, 1997, McGraw Hill

LabVIEW Power Programming, Gary W. Johnson, 1998, McGraw Hill

National Instruments Developer Zone, http://zone.ni.com

National Instruments LabVIEW Zone, http://www.ni.com/devzone/lvzone

Shape Analysis and Classification, Luciano da Fontoura Costa and Roberto Marcondes Cesar Jr., 2001, CRC Press LLC

The Engineering Handbook, Richard C. Dorf, 2003, CRC Press LLC

The Image Processing Handbook (4th ed.), John C. Russ, 2002, CRC Press

ONLINE RESOURCES

On a General Theory of Connectivity in Image Analysis, Ulisses Braga-Neto and John Goutsias, Center for Imaging Science and Department of Electrical and Computer Engineering, The Johns Hopkins University, http://cis.jhu.edu/~ulisses/papers/01-nsip.pdf

OpenG.org, the largest Open Source LabVIEW™ development community, http://www.OpenG.org

The Info-LabVIEW Mailing List, http://www.info-labview.org

Pattern Recognition in One Dimension, http://www.math-2000.com

SearchVIEW.net, "The Starting Point for LabVIEW on the Net," http://www.searchview.net

The Find All VIs Project, David A. Moore, http://www.mooregoodideas.com

What is Telecentricity?, Jim Michalski, Edmund Optics, http://www.edmundoptics.com/techsupport/DisplayArticle.cfm?articleid=261

COMPLETELY IRRELEVANT RESOURCES

Calvin and Hobbes, Bill Watterson, http://www.ucomics.com/calvinandhobbes

Industrious 2001, Mono*Crafts, http://www.yugop.com/ver3/stuff/03/fla.html

Melvin Kaminsky's biography, "I cut my finger: that's tragedy. A man walks into an open sewer and dies: that's comedy!" http://us.imdb.com/M/person-biography?Mel+Brooks

Tooheys Old Black Ale,
  http://www.australianbeers.com/beers/ tooheys_old/tooheys_old.htm
When in the USA, the author prefers Shiner Bock, http://www.shiner.com

Index